# Options for Initial Fuelling

Raj Raman
Robert Murphy
Laurence Smallman
John F. Schank
John Birkler
James Chiesa

Prepared for the
United Kingdom's Ministry of Defence

RAND EUROPE

The research described in this report was prepared for the United Kingdom's Ministry of Defence. The research was conducted jointly in RAND Europe and the RAND National Security Research Division.

**Library of Congress Cataloging-in-Publication Data**

The United Kingdom's nuclear submarine industrial base.
       p. cm.
       "MG-326/3."
       Includes bibliographical references.
       ISBN 0-8330-3784-6 (pbk. vol 3)
       1. Nuclear submarines—Great Britain—Design and construction. 2.
   Shipbuilding industry—Great Britain. 3. Military-industrial complex—Great
   Britain. 4. Defense industries—Great Britain.  I. Schank, John F. (John Frederic),
   1946– II. Raman, Raj. III. Title.

   V859.G7.U55 2005
   359.9'3834'0941—dc22

                                                                    2005010735

The RAND Corporation is a nonprofit research organization providing objective analysis and effective solutions that address the challenges facing the public and private sectors around the world. RAND's publications do not necessarily reflect the opinions of its research clients and sponsors.

**RAND**® is a registered trademark.

*Cover design by Peter Soriano*

*Photo courtesy of DML*

Published 2005 by the RAND Corporation
1776 Main Street, P.O. Box 2138, Santa Monica, CA 90407-2138
1200 South Hayes Street, Arlington, VA 22202-5050
201 North Craig Street, Suite 202, Pittsburgh, PA 15213-1516
RAND URL: http://www.rand.org/
To order RAND documents or to obtain additional information, contact
Distribution Services: Telephone: (310) 451-7002;
Fax: (310) 451-6915; Email: order@rand.org

# Preface

Several recent trends have warranted concern about the future vitality of the United Kingdom's submarine industrial base. Force structure reductions and budget constraints have led to long intervals between design efforts for new classes and low production rates. Demands for new submarines have not considered industrial base efficiencies resulting in periods of feast or famine for the organisations that support submarine construction. In an effort to reduce costs, government policies over the past decade have brought about a reduction in the submarine design and management resources within the Ministry of Defence (MOD). Yet the aforementioned production inefficiencies and increased nuclear oversight have resulted in greater costs.

Concerned about the future health of the submarine industrial base, the MOD asked RAND Europe to examine the following four issues:

- What actions should be taken to maintain nuclear submarine design capabilities?
- How should nuclear submarine production be scheduled for efficient use of the industrial base?
- What MOD capabilities are required to effectively manage and support nuclear submarine programmes?
- Where should nuclear fuelling occur to minimise cost and schedule risks?

This report addresses the last of those issues. The following companion reports address the first three issues:

- *The United Kingdom's Nuclear Submarine Industrial Base, Volume 1: Sustaining Design and Production Resources*, MG-326/1-MOD
- *The United Kingdom's Nuclear Submarine Industrial Base, Volume 2: MOD Roles and Required Technical Resources*, MG-326/2-MOD (forthcoming).

This report should be of special interest not only to the Defence Procurement Agency and to other parts of the Ministry of Defence, but also to service and defence agency managers and policymakers involved in weapon system acquisition on both sides of the Atlantic. It should also be of interest to shipbuilding industry executives in the United Kingdom. This research was undertaken for the MOD's Attack Submarines Integrated Project Team jointly by RAND Europe and the International Security and Defense Policy Center of the RAND National Security Research Division, which conducts research for the US Department of Defense, allied foreign governments, the intelligence community, and foundations.

For more information on RAND Europe, contact the president, Martin van der Mandele. He can be reached by email at mandele@rand.org; by phone at +31 71 524 5151; or by mail at RAND Europe, Newtonweg 1, 2333 CP Leiden, The Netherlands. For more information on the International Security and Defense Policy Center, contact the director, Jim Dobbins. He can be reached by email at James_Dobbins@rand.org; by phone at (310) 393-0411, extension 5134; or by mail at The RAND Corporation, 1200 South Hayes St., Arlington, VA 22202-5050 USA. More information about RAND is available at www.rand.org.

# Contents

# Tables

# Summary

In the United Kingdom, newly built submarines are fuelled where they are constructed: at the Barrow-in-Furness shipyard owned by BAE Systems. Devonport Management Limited (DML) currently refuels existing submarines once their initial fuel load is depleted[1] (and defuels them at retirement). Sustaining separate fuelling and refuelling sites has meant sustaining two sets of nuclear fuel–handling licenses. This has proven increasingly costly in a regulatory regime that expects to see continuous improvement. Such cost increases have led the Ministry of Defence (MOD) to consider the possibility of consolidating its nuclear fuel-handling capabilities at the existing re-fuelling site at DML.[2] If new submarines were fuelled at the DML dockyard, the Barrow yard could relinquish its nuclear fuel–handling license. Such a move could reduce expenditures on the current Astute-class attack submarine acquisition programme and future nuclear submarine projects.

Consolidation would, however, have complex implications for the Astute programme's cost and schedule. The MOD's Attack Submarines Integrated Project Team (ASM-IPT) thus asked the RAND

---

[1] Beginning with the Astute class, future submarines are planned to be fuelled for life and therefore do not need refuelling at mid-life.

[2] Nuclear refuelling activities for submarines have already been consolidated at DML. BAE Systems does not have the infrastructure or the regulatory approval to conduct refuelling at its Barrow-in-Furness shipyard. From a regulatory cost standpoint, refuelling is considerably more expensive than initial fuelling because of the nuclear hazards and consequences involved in handling spent, irradiated nuclear fuel compared with the lower risks associated with new, unused fuel.

Corporation to conduct an objective analysis of the full range of potential impacts that could result from consolidating dockyard nuclear fuel–handling capabilities at DML. The fuelling-refuelling consolidation problem was split into five components for analysis:

- Challenges to transporting an unfuelled submarine out of Barrow, through the Irish Sea, and on to Devonport
- Availability of DML facilities and the level of investment needed to make them suitable for fuelling the Astute boats
- Allocation of work between the two sites to produce an operational submarine
- Nuclear regulatory challenges at the two sites
- Contractual challenges that consolidation would have to meet, as well as public perceptions of consolidation-related impacts.

We assessed the consolidation-related savings (or costs) and the schedule risk (or potential for programme delay) associated with each of these sets of issues for three consolidation cases:[3]

1. Fuelling all Astute-class boats at DML
2. Fuelling Astute 1 at Barrow and all subsequent boats at DML
3. Fuelling the first three Astute-class boats at Barrow and all the others at DML.

Case 2 is based on the recognition that, given the first boat's advanced stage of construction, arranging for fuelling that boat at DML could cause delays. Case 3 is based on the possibility that the arrangements associated with consolidation would be easier to effect under a contract different to the current one, which covers the first three boats.

---

[3] These are compared with the current baseline strategy of fuelling all new boats at Barrow. Late in this project, after interaction with us regarding the DML fuelling alternatives, BAE Systems proposed a variant of the current baseline strategy (see "A New Proposal" below).

## Transportation

The most problematic segment of the transportation route is the exit from Barrow-in-Furness. The passage from the dockyard to the open sea is through the twisting Walney Channel, over 7 nautical miles long. Typically there is not enough water in the channel to keep a submarine off the bottom long enough to make the exit. A high tide is required, and not all high tides are sufficient. Depending on the precise draught of the boat, it may be necessary to wait three weeks or more for a sufficient tide. Even then, the tide runs so rapidly in and out of the channel that the speed of exit becomes a key factor. A fuelled boat under its own power can move down the channel quickly enough to exit in one tide. However, if the boat has not been fuelled and is, for example, towed out of the channel, two consecutive high tides would be required for exit. A deepwater staging point would thus have to be dredged midway down the channel. More time (and more opportunities) would, however, be available for the tow if the submarine were placed on a floating transport cradle, a pontoon-like device that would effectively reduce the submarine's draught. Less depth of water would then be required for exit. That is the arrangement we assume in the remainder of our analysis. It permits easier, less risky navigation than towing the boat without a cradle and would not cost as much as constructing the staging point required for a full-draught tow (£9 million vs. £20 million more than the baseline). There would also be more opportunities for exit than would be the case for either a fuelled submarine or a full-draught tow, so there would be less chance of delay.

Regardless of the exit option chosen, it would be most prudent if the MOD and BAE Systems reviewed the efforts leading to preparation of an Astute exit plan. A conservative plan similar to the one used for the Vanguard class in the past could result in unnecessary exit delays, even for a fuelled submarine. Safety can be assured through a more flexible plan that allows for the capabilities of modern technology to make precise sonar soundings of channel depth, predict swells at the channel mouth, and indicate the boat's position.

However, work must begin promptly on a flexible plan if it is to be in place by the time the first of class is ready to depart.

## Facilities

Facilities at DML are used to support in-service submarines. While there are periods during which a new boat could be fuelled, their occurrence is not convenient to the Astute programme schedule. The first of class either would have to wait many months or would have to be launched without some components that are most efficiently installed when the hull is still open. That installation would then have to be performed within the closed hull at DML, or the hull would have to be reopened there, both of which are costly and impractical alternatives.

Initial fuellings beyond the first of class may encounter conflicts with other submarines requiring docking periods for maintenance, but any delays on that account are likely to be minimal. Thus, only case 1 is seriously affected. For all cases, facility investments specific to fuelling Astute-class boats are difficult to judge. Although DML claims that these investments would be minimal, uncertainty will remain without having a detailed plan.

## Workload

More production labour would not be required to fuel the boats at Devonport than to do so at Barrow under the fuelling procedures now in place. However, the work would be allocated differently, which has implications for schedule risk on the fuelling plan at Barrow. Additional oversight would be required if fuelling activities were consolidated at DML. This would entail DML personnel overseeing the manufacturing process at Barrow to ensure the submarine they fuel has met manufacturing standards, and BAE Systems personnel would serve as the design authority overseeing the fuelling and subsequent testing and commissioning at DML.

If fuelling were to occur at DML, construction work that would have been done in parallel with fuelling at Barrow would have to be finished at DML. Thus, construction processes required before launch and conducted in parallel with the core load would now be conducted in series with the core load. As a result, the construction schedule would be extended by the 15 weeks for the core load at DML, plus a few weeks for transport and preparing for fuelling.

This is an underestimate of the delay beyond the first two boats. By the time the third boat is built, BAE Systems expects to be taking a different approach to fuelling that is expected to save time (about one month) and labour (about 160,000 man-hours per submarine, subject to confirmation of feasibility of the construction change and satisfaction of security issues). Because, in that case, fuelling would become an integral part of constructing the hull, this approach could not be taken if fuelling were done at DML. Thus, relative to the Astute 3 baseline schedule, fuelling at DML would take the five months extra as for the first of class, plus the additional month. This applies to all three cases under consideration.

If a vessel is fuelled at DML, both BAE Systems and DML would need to have people stationed at each other's site. BAE Systems personnel would be required for insights and input into post-launch activities, and DML personnel would want to monitor at least those aspects of construction associated with the nuclear steam-raising plant. This oversight workload is not required in the current plan to fuel at Barrow and would cost approximately £2.2 million per boat.

## Nuclear Regulation

Two agencies regulate nuclear activities at the Barrow-in-Furness shipyard and DML: the MOD's Naval Nuclear Regulatory Panel (NNRP) and the Health and Safety Executive's Nuclear Installations Inspectorate (NII). At the time the last submarine was fuelled at Barrow, the NNRP's jurisdiction was restricted to the submarine and the NII's to the rest of the site. In 1996, the two organisations signed an agreement whereby the NII would attend also to the interface between the submarine and the surrounding site, and there would be greater joint attention to all nuclear-related activities on site. The result has been, from the regulator's viewpoint, a recognition that a more modern approach to site safety cases and a greater coordination between operating organisations and regulatory regimes are now needed. Addressing this issue has seen a large increase in costs, which has been regarded by the MOD's contractors as directly attributable to regulatory compliance, although the basic compliance requirements are long standing.

From the contractor's viewpoint, the nonprescriptive environment fosters uncertainty in scope of work leading to cost growth. As a result, our own discussion of regulatory compliance costs, which draws from contractor perspectives and data, also has to be caveated with an element of uncertainty.

About £20 million has already been spent at Barrow in this connection and the expenditure of another £100 million in recurring and nonrecurring expenses is anticipated for the first three boats. Of that amount, £30 million is allocated for maintaining manufacturing quality assurance standards sufficient for a nuclear-powered submarine, which will have to be done regardless of where the submarine is fuelled. The other £70 million is associated with fuelling and represents a potential saving at Barrow if the boat is fuelled at DML. The compliance activities are, however, under way, so the longer a decision to fuel at DML is postponed, the more will be spent and the less will be available for saving. Figure S.1 shows the potential savings remaining over the next eight years. The break in the curve indicates that by 2006, the fuel for the first of class will have arrived at Barrow and no further savings can be realised (i.e., the costs must accrue) until the boat leaves Barrow. Subsequent compliance savings (about £21 million) can still be realised in case 2, in which the second boat (and each one following) is fuelled at DML. None is realised in case 3, in which fuelling at DML awaits the next contract. In all cases, however, the Barrow yard would save £7.3 million per year in recurring nuclear regulatory compliance costs and nuclear-related overhead

**Figure S.1**
**Available Savings Associated with Site Licensing at Barrow Reduce Annually**

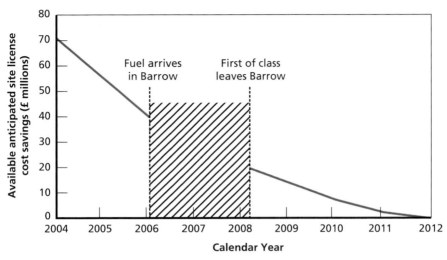

for fuelling Astute 4 and beyond at DML, once the Barrow shipyard is decommissioned as a nuclear fuel–handling site for a one-time cost of £15 million.

Costs at DML have not been determined, but some fuelling-related activities and assets are already covered as part of the more stringently regulated refuelling and defuelling already done there. Nevertheless, a safety case would be required for the new fuelling process. It may take some time to satisfy regulators that DML can take full responsibility for fuelling, given its lack of involvement in the Astute programme to date. Presumably, regulatory issues will have been resolved by the time that boats under the next contract are to be fuelled; therefore, case 3 would not be affected.

It is important to note that the baseline programme, in which all Astutes are fuelled and delivered from Barrow, may itself be delayed by satisfying nuclear regulatory demands. The regulatory work left on hull, and the procurement of a floating transport cradle. For cases 1 and 2, these costs still accrue to later boats, but the tally is dominated by the large savings in safety case preparation and other nonrecurring nuclear regulatory compliance costs accruing to the current contract at Barrow. These savings are considerably larger for case 1 than for case 2. It should be kept in mind that important elements of these savings are based on data from a single source, which we have not submitted to a thorough critical review. However, the difference between the savings for case 1 and those for the other cases appears unlikely to be reversed on closer examination.

Programme delays would total 14 months for the first of class, because of the need to wait for facility renovation at DML and the need to run fuelling and parallel processes at Barrow in series when shifting to DML (see Table S.2). That is a conservative estimate; it omits potential further delays related to nuclear regulatory demands and contractual matters that are difficult to quantify. Of these, only the running of parallel processes in series applies to subsequent boats, so the anticipated delays would be six months from currently planned delivery dates. Some further delay for contractual issues is possible for the second boat. Note that, in the table, a BAE Systems action item is prescribed to avert the possibility of schedule delays from failing to move promptly towards a flexible plan for moving ships from the Barrow dockyard to the open sea.

**Table S.2**
**Schedule Risks from Fuelling New Submarines at DML**

| | Transportation | Facilities | Workload Impact | Nuclear Regulatory Issues | Contractual Issues |
|---|---|---|---|---|---|
| Baseline | | No delay | No delay | Uncertain scope | No delay |
| Case 1 | BAE Systems action item | 8-month delay for Astute 1 | Up to 6-month delay | NII concerns | Significant delay |
| Case 2 | | Minimal delay | Up to 6-month delay | No delay | Some delay |
| Case 3 | | Minimal delay | Up to 6-month delay | No delay | No delay |

NOTE: See the list near the beginning of summary for the definition of each case.

In sum, case 1, for which the greatest savings accrue, would be severely affected (with respect to the first of class) by delays. Case 3 would be less affected by delays but realises no savings—and the potential for savings is what motivates the transfer of fuelling to DML. Case 2 affords modest savings and is subject to some delay. It is noteworthy, however, that we can account here only for currently anticipated costs. Should further tightening of the regulatory regime require future safety cases, those costs would be preempted at Barrow by consolidating nuclear fuel–handling at DML, and the savings from cases 2 and 3 could be dramatic.

## A New Proposal

Towards the end of this study, BAE Systems proposed a new approach to fuelling that it believes would result in substantial savings to the Astute programme without the need to consolidate all fuelling activities at DML. The proposal includes measures to reduce risks of handling nuclear fuel and provides that activities necessary to complete the testing of the submarine be done at either Barrow or Faslane. BAE Systems estimates a savings of £18 million NPV in the first case (Barrow) and £50 million in the second (Faslane), since the reduction in nuclear hazards associated with such an approach would simplify the preparation of safety cases. To these costs must be added the cost of transporting the submarines to Faslane, which we estimate at £12 million NPV for the eight Astute-class boats.

We cannot independently validate the BAE Systems estimates. Furthermore, important uncertainties remain regarding

- the availability of berths, services, or testing equipment necessary for initial Astute power range testing (PRT) at Faslane
- the effort required to prepare a safety case for initial Astute PRT for Faslane
- any potential limitations on the number of days of critical operations that can be performed at Faslane
- issues of ownership of the untested submarine

- availability of suitably qualified and experienced personnel.

Despite the uncertainties, this new proposal has to be regarded as promising. The savings achievable, if validated, are on the order of those attainable only through fuelling all Astute-class boats at DML, including the first of class, whose fuelling at DML would require a delivery delay of more than a year. BAE Systems claims no delays from planned delivery dates under its new proposal; further assessment is required to verify this claim. Meanwhile, BAE Systems is moving forward with the proposal.

## Recommendations

Based on the analysis summarised here, we recommend that the MOD

- not consider fuelling the first Astute-class boat at DML
- take prompt action in analysing the latest proposal submitted by BAE Systems to reduce nuclear consequences of hazards at Barrow.

If upon further analysis the late BAE Systems proposal is found unlikely to produce the savings and risk reduction anticipated, a decision could be made at that time for the MOD to engage with the regulators in assessing options. It should look in detail at relevant aspects of the build programme, support facilities and options, and conduct a more detailed feasibility study for cases 2 and 3. The MOD should also consider the possibility that future nuclear regulatory requirements and restrictions could make both cases 2 and 3 seem advantageous, even with respect to the latest BAE Systems proposal. There is also the possibility, of course, that the current plan of fuelling all new boats at Barrow will emerge as preferable. Regardless, the MOD and BAE Systems need to

- review promptly the transportation challenges associated with moving Astute from Barrow to the open sea, regardless of whether the boat is fuelled or unfuelled, and produce a flexible exit plan that minimises potential schedule risk.

# Acknowledgments

This research would not have been possible without the assistance of several key individuals within the Ministry of Defence; the nuclear submarine industry, including BAE Systems, DML, and Rolls-Royce; the Nuclear Installations Inspectorate and the Naval Nuclear Regulatory Panel; and the Naval Sea Systems Command's Nuclear Propulsion Directorate. Our sincere gratitude goes to the Attack Submarines Integrated Project Team's Muir Macdonald and Nick Hunt; we especially thank the ASM-IPT's Helen Wheatley for her valuable assistance in providing data and information and facilitating our interactions with multiple organisations throughout the course of this project.

Commander JJ Taylor, of the Submarines IPT, provided data on planned retirements and maintenance actions for the current fleet of submarines in the Royal Navy. Captain Steve Firth of the Warship Support Agency supplied valuable insights into the maintenance issues at DML. David Shaw and Maureen MacAlonan of the Ministry of Defence's Price Forecasting Group provided important cost data for our research. Commodore Andrew McFarlane of the NNRP, Commodore Michael Bowker of the Nuclear Propulsion IPT, Commander Barry Tarr of the Attack Submarines IPT, and Robbie Gray of the NII provided critical information on nuclear regulatory issues. Morgyn Davies and Stephen Quinn of the MOD Salvage and Mooring Operations IPT and Stephen Young and David Carpenter of the Associated British Ports, Barrow-in-Furness, provided specialist maritime transportation advice.

This list would not be complete without acknowledging the support of several key individuals within the nuclear submarine industrial community. Huw James, Duncan Scott, Mark Dixon, Brian Devenny, and Michael Wear from BAE Systems; Mike Owen, Peter Whitehouse, and John Knox from DML; and Steve Ludlam from Rolls-Royce Naval Marine—all provided critical pieces of information related to the challenges of fuelling Astute-class submarines at DML. The overall perspective provided by Carl Oosterman and Steve Trautman of the US Naval Sea Systems Command (NAVSEA 08) was extremely helpful in addressing the important issues in such a complex problem. Finally, we thank our reviewers Giles Smith of RAND and David A. Griffiths of SERCO Assurance UK for their valuable suggestions and insights.

At RAND, Deborah Peetz provided her typically excellent support to the overall research, and Phillip Wirtz edited the final document.

The abovementioned individuals helped us with functional information and suggested some implications. We, however, are solely responsible for the interpretation of the information and the judgements and conclusions drawn. And, of course, we alone are responsible for any errors.

# Abbreviations

| | |
|---|---|
| ASM-IPT | Attack Submarines Integrated Project Team |
| CNNRP | Chairman, Naval Nuclear Regulatory Panel |
| CSALMO | Chief Salvage and Mooring Officer |
| DLO | Defence Logistics Organisation |
| DML | Devonport Management Limited |
| HSE | Health and Safety Executive |
| IPT | Integrated Project Team |
| MOD | Ministry of Defence |
| NII | Nuclear Installations Inspectorate |
| NNRP | Naval Nuclear Regulatory Panel |
| NPV | net present value |
| NSRP | nuclear steam-raising plant |
| PRT | power range testing |
| SSBN | nuclear-powered fleet ballistic missile submarine (ship submersible ballistic nuclear) |
| SSN | nuclear-powered attack submarine (ship submersible nuclear) |

# CHAPTER ONE

# Introduction

In the United Kingdom, newly built submarines are fuelled[1] where they are constructed: at the Barrow-in-Furness shipyard owned by BAE Systems. Devonport Management Limited (DML) currently refuels submarines once their initial fuel load is depleted. Sustaining separate fuelling and refuelling sites has meant maintaining two sets of nuclear licenses. This has proven increasingly costly in a regulatory regime that has been growing more demanding. A facilities upgrading project at DML undertaken partly to improve compliance against current standards and hence reduce risk to the public overran its budget, and near-future compliance estimates for the Barrow yard have sharply increased. Such costs have led the Ministry of Defence (MOD) to consider the possibility of consolidating its nuclear fuel–handling capabilities at the existing refuelling site at DML. This would require Barrow to relinquish its nuclear fuel–handling capabilities with new submarines fuelled at the DML dockyard.

Consolidation would have complex implications for the Astute program's cost and schedule. The MOD's Attack Submarines Integrated Project Team (ASM-IPT) thus asked the RAND Corporation to conduct an objective analysis of the full range of potential impacts that result from consolidating dockyard nuclear fuel–handling capabilities at DML.

---

[1] Throughout this report, *fuelled, fuelling*, etc., will refer only to activities performed on a newly constructed submarine. Subsequent replenishment of fuel will be indicated by *re-fuelled, refuelling*, etc.

## Background

Nuclear regulations play a key role in the manufacture and maintenance of nuclear submarines in the United Kingdom. The MOD's Naval Nuclear Regulatory Panel (NNRP) and the Health and Safety Executive's Nuclear Installations Inspectorate (NII) are the two organisations that provide regulatory oversight. Until Astute, the NNRP's main area of responsibility had been the physical boundaries of the submarine, while the NII's main area had been the safety of operations within the manufacturing or maintenance site. In other words, the submarine with its complex machinery was treated as a 'bubble' directly under the authority of the NNRP, while the NII had responsibility for operations at the site. In 1996, the MOD agreed with the NII to provide the NII site licensee with data on the design of the submarine nuclear reactor. The intent was not for the NII to look into the reactor design but instead for it to gain understanding of the reactor's safety-related matters and how they interacted with the shore-based facilities. As a result, the bubble disappeared and the NII's responsibility grew to include the nuclear activities related to the submarine reactor while it was on the manufacturing or maintenance site.

Although the nuclear regulations have not changed, the 1996 agreement has caused a change in the way the regulators operate. More specifically, the NII has to be convinced that all operations related to the nuclear reactor are being handled in a safe manner while the reactor is in the manufacturing or maintenance site. This has resulted in the site licensees' experiencing a sharp increase in the complexity of safety cases required to convince the NII that their products and processes meet requisite safety standards. Such an effort consumes a substantial amount of resources, ultimately borne by the MOD.

The first impact of the 1996 agreement was seen at DML. Beginning in 1997, DML upgraded various facilities while addressing the changes in the regulatory environment. The practical challenges and subsequent cost effects of how the nuclear regulation regime would affect this project were not fully appreciated by any of the parties involved.[2] The imperative to meet milestones to support the submarine refit programme also contributed to cost increases, and the facilities' design evolved to take account of the regulators' observations, requiring additional construction and work.

The only other dockyard site handling nuclear fuel is the Barrow-in-Furness shipyard. After a hiatus of almost a decade since the last Vanguard-class nuclear submarine was built at Barrow, the site is currently in the process of preparing safety cases for fuelling and critical operation of Astute-class submarine reactors. BAE Systems had originally projected a cost of approximately £20 million for this effort, which was based on the regulatory environment in which the Vanguard class was manufactured. Changes in the environment since then have resulted in an increase in anticipated site licensing costs by an additional £100 million over the original proposal.

## Analytic Approach

The fuelling-refuelling consolidation problem was split into five key components for analysis:

- Challenges to transporting an unfuelled submarine out of Barrow, through the Irish Sea, and on to Devonport
- Availability of DML facilities (e.g., docks) and the level of investment needed to make them suitable for fuelling the Astute boats
- Allocation of work between the two sites to produce an operational submarine
- Nuclear regulatory challenges at the two sites
- Contractual challenges that consolidation would have to meet, as well as public perceptions of consolidation-related impacts.

---

[2] UK National Audit Office, *The Construction of Nuclear Submarine Facilities at Devonport*, Report by the Comptroller and Auditor General, HC 90 Session 2002–2003, 6 December 2002.

In each of these aspects of the problem, we were interested in the implications for Astute programme cost and for the risk of schedule delay. (Except where indicated otherwise, the cost and schedule analyses presented in this volume are based on information available as of summer 2004.)

Cost estimates were obtained from multiple sources as shown in Table 1.1 and crosschecked for their validity and reasonableness. The A2B Fuelling Study was conducted in 2002 by BAE Systems Astute Class Limited. Its goal was to establish the cost effectiveness of fuelling Astute 2nd Buy (A2B—i.e., Astute 4 and onwards) at DML rather than the Barrow-in-Furness yard (operated by BAE Systems Submarine Division). The study was based on limited cost information readily available and non-verifiable assumptions based on Defence Procurement Agency and Astute prime contractor office in-house experience. The report was completed without any direct approach to either DML or BAE Systems Submarines. Apparently, then, the costs reported by the study were unverified by the two main parties involved. To obtain a more robust estimate, we contacted these organisations and several others as listed in the table to understand the costs related to fuelling at DML, beginning with the first of class. Rough order-of-magnitude estimates were provided by the different sources based on their expertise in the five different areas.

**Table 1.1**
**Sources of Cost Estimates for the Five Areas Impacted by Initial Fuelling at DML**

|  | Transportation | Facilities | Workload Impact | Nuclear Regulations | Contractual |
|---|---|---|---|---|---|
| A2B Fuelling Study | X |  | X |  |  |
| BAE Systems | X | X | X | X | X |
| DML |  | X | X | X | X |
| Rolls-Royce |  |  | X | X |  |
| MOD Pricing and Forecasting Group |  |  |  | X |  |
| CSALMO | X | X |  |  |  |

Information on potential schedule effects, various technical options for cost and risk reduction, and qualitative considerations came from a literature review and numerous interviews. Those interviewed included personnel from the ASM-IPT, the MOD's Pricing and Forecasting Group, the Chief Salvage and Mooring Officer (CSALMO), Associated British Ports Barrow-in-Furness, the NII, the NNRP, BAE Systems, DML, and Rolls-Royce Naval Marine (the nuclear plant contractor).

The central issues in this study could be resolved by comparing cost and schedule risk of the current plan to fuel new boats at Barrow with those of the alternative of fuelling at DML. However, because Astute 1 construction is already well under way, and work has already begun on the next two boats, the question arose as to whether consolidation of fuelling at DML might be phased in. We thus defined three cases to be considered against the 'baseline' of maintaining the status quo, that is, fuelling all new boats at Barrow (see Figure 1.1).

**Figure 1.1**
**Different Cases of Fuelling New Boats at DML**

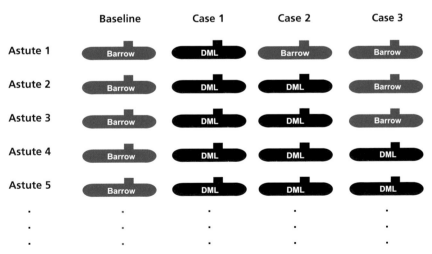

now in the fleet could cost relatively more to maintain due to their age when compared with the new Astute class. However, a recent decision calls for a reduction in the number of SSNs in the fleet from the current eleven boats to eight. There is some slack between the current scheduled delivery of the first Astute boat and the date needed to maintain eight SSNs. Also, because the delays associated with the different fuelling options turn out not to be lengthy, it is unlikely that further maintenance actions will be taken on boats about to retire, so the cost of any delay is likely to be small, and we do not estimate it. We also do not address such intangible costs as delays in the introduction of a more effective and efficient submarine to the Royal Navy for some period of time.

## Organisation

This report addresses different aspects of the problem in six separate chapters followed by our research conclusions and recommendations in Chapter Eight. We begin with Chapter Two, which addresses the technical challenges and cost of transporting an unfuelled submarine out of Barrow into the Irish Sea and on to the Devonport dockyard. Chapter Three addresses the facility requirements at DML; it looks at the availability of docks at DML along with the investment required to make the facilities suitable for the initial fuelling activity there. Chapter Four addresses the cost and schedule impact on workload related to the initial fuelling effort; it compares the option of fuelling at DML against the baseline of conducting all the work at Barrow. Chapter Five addresses the issues related to the change in nuclear regulatory environment and how it impacts the cost and schedule risks associated with the problem. Chapter Six addresses the contractual challenges and public perception issues that need to be overcome to make such a move possible.

Net present value analysis is conducted to compare the cost of the three cases against the baseline, and a comparison of overall schedule risk is also made. Recent proposals made by BAE Systems to reduce nuclear hazards and resulting consequences at Barrow are addressed towards the very end in Chapter Seven. This latest proposal was not subjected to a detailed analysis because of the lack of time and resources available. Nevertheless, we feel it is worth mentioning the proposal in this document because of its potential of providing significant savings while reducing the possible nuclear consequences of hazards at Barrow, thereby influencing the decision to fuel at DML.

# CHAPTER TWO

# Transportation

Fuelling at DML entails transporting the unfuelled submarine from Barrow. An unfuelled submarine would obviously need assistance to move from the BAE Systems shipyard at Barrow-in-Furness to the DML facilities at Devonport in Plymouth where it would be fuelled. We assessed the feasibility of such a move by analysing several aspects of the problem:

- What are the challenges in the transportation route from Barrow-in-Furness to Devonport?
- What are the different transportation options for such a move?
- What are the navigational and schedule risks and costs of the transportation options considered?
- How does the navigational and nuclear regulatory environment affect such a move?

In this chapter, we address these issues in the order listed above. We conclude with decisions the MOD and BAE Systems need to make soon if the Astute programme schedule is to be met.

## Transportation Route

Masters of vessels in UK waters are required to follow regulations for safe navigation.[1] 'Safe navigation' is an all-encompassing term that places significant responsibility on the master and involves all aspects of the movement of the vessel so that the ship and her crew, as well as other users of the water, are not jeopardised. The regulations are comprehensive and give specific instruction and guidance.[2] As discussed in more detail later, our work identified the principal concerns for the transportation options as grounding on a submerged hazard or seabed, or collision with another vessel or fixed structure. The risk of grounding is reduced when the width and depth of a navigable channel are greatest. In tidal waters, this means that the greatest period of safe navigation is likely around the time of high water (see Figure 2.1 for definition of terms). The length of this period will be related to the height of tide and the draught of the vessel as well as the distance to safe water and the speed of advance of the vessel. The tidal stream is an important factor in confined waters too. Vessels that are large in relation to the available depth and width of the navigable channel, or that are restricted in their manoeuvrability or ability to react to unforeseen circumstances, are particularly vulnerable in fast tidal streams. A large submarine using its own power would fall into the first category, while both categories (i.e., large and restricted manoeuvrability) would characterise a towed submarine.

In open waters it is possible to choose a route that avoids dangerous shallow areas, which reduces the risk of grounding or collision with submerged hazards. The presence of other maritime traffic and the potential impact of severe weather become the more significant

---

[1] The regulations of the International Convention for the Safety of Life at Sea (SOLAS V) were implemented under UK legislation by the Merchant Shipping (Safety of Navigation) Regulations 2002.

[2] The MCA publishes the UK regulations online at mcanet.mcga.gov.uk/public/c4/regulations/safetyofnavigation/index.htm.

**Figure 2.1**
**Tide Terminology**

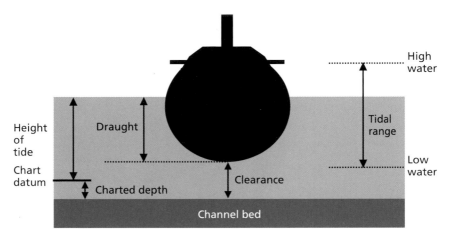

RAND *MG326/3-2.1*

hazards. Manoeuvrability is an important factor in avoiding collisions with other vessels. Although other vessels must yield the 'right of way' to vessels in a restrictive towing arrangement, the difficulty of the tow, its slow speed, and the time needed to alter course could add to the potential risk of collision. Once at sea, weather avoidance for slow vessels, such as a towed submarine, is more difficult and more important than it is for faster vessels or those able to manoeuvre independently.

**Description of Route**

There are three phases involved in the movement of an unfuelled submarine from Barrow-in-Furness to Devonport. Exit from Barrow makes up the first phase of the transit, followed by the deep Irish Sea in the second phase, and finally entry into Devonport in the third phase. We looked carefully at the navigation factors that accompany each phase to understand better the potential restrictions that might affect safe transportation of Astute.

passing from the still waters of the dock into the fast-moving tidal stream in the channel would be very difficult and possibly dangerous for the submarine and attendant tugs.[3] Towards the southern end of Walney Channel, as the channel approaches open water and the Irish Sea, there is less protection from the prevailing southwesterly winds and swell. Winter squalls, summer haze, or mist and fog can reduce visibility throughout the year.

The ability of any vessel to navigate safely in Walney Channel depends on the clearance around the vessel; large vessels with deep draughts, the category that applies to submarines, suffer significant departure restrictions. Vessels with very broad beams could be affected, too, but this restriction does not apply to the Astute. A further consideration is the combination of the rate of change of the height of tide and the distance to deeper water. In most circumstances, a vessel with a deep draught will need to complete the passage during one high tide if it is to avoid being grounded. In fact, departures for submarines down the Walney Channel are limited to the highest of high tides, since only these provide sufficient water along the length of the channel for the duration of the exit. Even then, a submarine leaving the dock system has to race the tide out to safer water. Finally, the turns need to be completed precisely to ensure the submarine remains in safe water.

**Open-Sea Passage.** The Irish Sea is busy with ferries crossing east or west between Great Britain and Ireland, slow-moving fishing vessels that often group together, and vessels on passage north or south to other UK ports or the Atlantic. A towed vessel and associated tugs, where the nature of the tow is restrictive, could expect the majority of other vessels to take avoiding action when necessary. This does not apply in all circumstances, however, and the tug and tow would need to manoeuvre occasionally, particularly for fishing vessels

---

[3] The tidal window before high water is specified in the Vanguard Exit Plan that we refer to later in this section. Since the completion of our research, a more technical assessment of the impact of the tidal stream effects has been commissioned. This is likely to give better information than that available to us, and other factors, including the underkeel clearance across the Ramsden Dock Basin cill, could redefine the limiting conditions for departure from the dock system into Walney Channel.

engaged in fishing. However, a submarine fitted with appropriate towing points and attended by the required number of tugs should present no hazard.

In the Irish Sea and particularly around Land's End, adverse weather could impede a tow, and in extreme conditions, especially if there were strong winds from the west and a large Atlantic swell, the passage would be too dangerous to attempt. There would be a significant danger of being forced onto the coast. Such stormy weather is easy to predict, however, and the danger of such an occurrence would be known before having to leave the safety of Barrow-in-Furness.

**Devonport Entry.** The Queen's Harbourmaster controls Devonport and Plymouth Sound, and for submarine movements all other traffic is restricted to reduce the potential hazards and ease navigation. There are deepwater buoys immediately inside the Western Channel entry to Plymouth Sound that are regularly used by submarines, and one of these could be used while an unfuelled Astute waits for a suitable high tide to complete the entry to Devonport. There is sufficient depth of water on most tides and at most times such that there would be few restrictions for an arriving submarine. In addition, tidal streams are unlikely to be problematic at this point.

The most significant danger for arriving vessels is the tidal stream in the vicinity of Vanguard Bank, farther into the entry, and the approach to this sharp turn is crucial, although for a towed vessel assisted by tugs it is easily achieved by experienced admiralty pilots. The route is deep enough for the majority of vessels to use so that the time of rounding Vanguard Bank can be adjusted to minimise the effects of the tidal stream. Deep-draught vessels would not have the flexibility to enter on low-water spring tides, and slow-moving vessels would have to avoid periods of the fastest flowing tidal streams.

The Devonport Entry does not require a vessel to race against a rising or falling tide, since there is sufficient depth at all stages once past the initial shallow area in Plymouth Sound. Plymouth Sound is generally protected from all weathers, and only the strongest winds would impose movement restrictions.

**Comparative Assessment.** Clearly, the Barrow Exit is the transit's most difficult phase. The departure restrictions from the dock system, the large tidal range, the distance to safe water, and the draught of the submarine are major factors that have to be addressed in order to plan a safe departure for a fuelled or unfuelled Astute. Additionally, the likely slow speed of a towed unfuelled submarine would make that option tenuous in a single high-water cycle. Finally, it appears that in any conditions that would allow for a safe departure from Barrow-in-Furness, very shortly after starting down Walney Channel the point of no return would be reached and there would be very few if any options to reach or return to safe water in the event of an emergency. Conversely, the factors affecting safe navigation of the Irish Sea Passage and Devonport Entry are manageable by prudent use of established navigational procedures; each of these phases allows for a flexible response to an unexpected event. Given the challenges of the Barrow Exit, we examine it here in more detail.

**Analysis of Barrow Departure**

Working from admiralty charts and tidal data, we investigated the implications of draught and speed for exit opportunities out of Barrow. The weight of the fuel has minimal impact on the draught of a submarine, and the draught of an unfuelled Astute is comparable to that of a fuelled one. As a result, the following analysis is applicable to both. To avoid raising the classification of this report, we do not use specific draughts for either the Astute or Vanguard classes of submarine. To compare different tides directly, we assume no tidal stream. The variations in strength and direction of the tidal stream will need to be calculated or measured for detailed exit planning and safe navigation. Nor are we able to quantify the effects of speed in shallow water, a condition known as *squat*, for the Astute class. Qualitatively, though, it is an important limiting factor on the speed that is available for use by a vessel with limited underkeel clearance.

    **Speed and Draught.** From the time of high water, a vessel with a shallower draught will remain afloat for longer than one with a deeper draught. For a vessel moving along a channel when the water is receding, the further variable of speed is introduced. A faster vessel will complete the passage more quickly than one that is slower. Thus, greater speed can make up for greater draught, because a vessel with enough draught to risk grounding before exit is completed could, if it were faster, exit the channel before the receding waters presented the risk. For a given draught, then, there is some minimum speed that will safely get the boat out of the channel. By choosing one high tide and fixing the start time to high water, we are able to use tide tables to evaluate the relationship between draught and speed (see Figure 2.4).

Figure 2.4 shows the relationship between a vessel's draught and the consequent average speed it needs to complete the exit from Ramsden Dock to safe water. Submarines typically have deep draughts in the region of 10 metres, where the curve shows that for small changes in draught there are consequent large changes in the minimum average speed required. This has important implications for towing options. For example, if the draught of the submarine

**Figure 2.4**
**Relationship Between Draught and Minimum Speed for Safe Departure from Barrow**

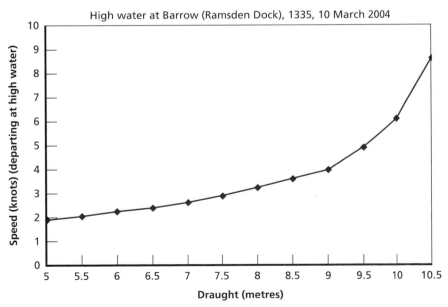

could be reduced, more time would be available to accommodate a lower speed. However, for some towing options, the submarine would not achieve a maximum speed of more than 4 knots, and the average, particularly where precise turns are required, might be closer to 3 knots. At those speeds, the channel cannot be navigated in one tide, and a mid-channel staging point would be required.

**Exit Opportunities and Draught.** The heights of high tide change throughout the month and across the year. Not all high tides at Barrow-in-Furness reach sufficient height to allow deep-draught vessels to complete the exit without grounding. For a given achievable speed, a shallower-draught boat will have more opportunities to exit over the course of a month or a year. Recall that the availability of the comparatively slack water constrains departure to between 45 and 30 minutes before high tide. The precise timing within that window also affects exit opportunities. Earlier departures have more time available at higher waters and thus allow more exit opportunities for a given draft.

These relationships are shown in Figure 2.5 for February 2004 at a fixed vessel speed representative of a fuelled submarine.[4] As can be seen, small reductions in draught lead to a dramatic increase in the number of exit opportunities available (out of a total of 56 high tides in the month shown). Drafts around 10 metres allow few opportunities, and environmental conditions will prevent some of these exits when, for example, visibility is poor or there are strong winds and significant swell.

It is also noteworthy that, particularly for a 10-metre draught, small delays in departure can significantly reduce exit opportunities. On any given day, factors such as weather, machinery breakdowns and the proficiency of the personnel involved can work with or against each other to affect the precise timing of departure, regardless of what the intentions may have been. Thus, a departure aimed for

---

[4] Appendix A presents a more detailed description of our analysis.

**Figure 2.5**
**Exit Opportunities for the Astute as a Function of Draught at a Fixed**
**Vessel Speed**

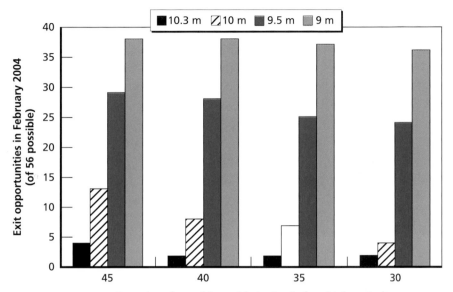

NOTE: 10.3 metres is the maximum draught of submarine that can be accommodated within the dock system.
RAND *MG326/3-2.5*

45 minutes before high tide might slip to 30 minutes, with a concomitant drop of 10 to 70 percent (depending on draught) in the probability that conditions will still be suitable for exit.

## Transportation Options

If newly built submarines are to be fuelled at Devonport, they will need assistance leaving Barrow. Based on interviews with marine salvage experts, the harbour master at Barrow, and those in the shipping industry, we considered three options for transportation:

- **Clean tow.** In this situation, the boat would be allowed to float unassisted at its design draught of about 10 metres and would be towed down the channel by tugs. As our analysis indicated, the reasonable maximum speed of 3 knots for a tow is considerably less than the minimum speed of about 6 knots required for the draught of a submarine, and a deepwater staging point would need to be constructed at some point midway along Walney Channel.[5] This would be needed to allow the submarine to wait in safe water for the next high tide. The submarine would be towed to Plymouth and into Devonport with tugs and towing arrangements that would allow for a stable tow.
- **Floating transport cradle.** This option could range from the simple addition of a suitably robust pontoon system to something more elaborate approaching the structural complexity of a floating dock. The exit plan design, any regulatory requirements, and marine engineering practice would determine the nature of the cradle. The concept takes advantage of the significant changes that occur for small reductions in the draught of the submarine; the design would allow an unfuelled Astute to depart in one tidal cycle. The submarine would be placed in the cradle in the dock system and released in the deeper water of Morecambe Bay prior to being towed to Devonport. Because exit opportunities would be much more frequent than with a clean tow, a few days' delay might be allowed if strong winds and swell made use of the cradle hazardous.
- **Heavy-lift ship.** In this option, the unfuelled submarine would be loaded on to a ballasted self-powered heavy-lift ship inside the dock system. The ship would then de-ballast to reduce its draught for departure. A floating transport cradle might be needed to load the submarine on to the ship, as the deep subma-

---

[5] Maintenance dredging is used to keep Walney Channel in its current condition. We were informed during our research that further extensive dredging would encounter significant environmental objections. A deepwater staging point would require less overall dredging (in terms of volume) than that needed to deepen the whole channel, and so we use this as the example here.

rine draught would impose design constraints on the heavy-lift vessel. There is a deep dive hole in the Buccleuch Dock that would allow a heavy-lift ship to submerge sufficiently to load an Astute, although the Ramsden dock gate widths would impose beam restrictions. This could mean that a purpose-built vessel would be required, which would be expensive. Alternatively, the cradle could be used to move the submarine to load at the deepwater berth in Walney Channel, or down Walney Channel as described above to deeper water, where the heavy-lift ship could then be loaded. The shallow draught and relatively high speed of the heavy-lift ship would allow maximum flexibility for departure, if used in the channel, and, in any event, the highest degree of manoeuvrability during the Irish Sea Passage. The cradle and submarine would be unloaded from the heavy-lift ship, and the submarine floated free prior to entry to Plymouth Sound, whence the submarine alone would be towed into Devonport.

## Assessment of Transportation Options

We assess the transportation options and compare them against navigational and schedule risks and cost. Risks are presented in terms of a qualitative ranking of the transportation options relative to each other. Nonrecurring and recurring cost estimates for each option are also presented.

### Navigational Risk

In our interviews with marine salvage experts and experienced mariners, concern was expressed about the manoeuvrability of any tug and towed submarine, especially during the turns in the outgoing tidal stream in Walney Channel. These would be especially difficult to manage with the clean tow option. Cutting the draught a little by using the floating transport cradle would increase the safety margins enough to substantially mitigate these concerns. Even so, the tow would require skilful handling. The heavy-lift ship is the least navigationally risky of the three options. (The assessment is summarised in Table 2.1).

## Schedule Risk

As discussed above, the risk of schedule delay will be less for a vessel that has a shallower draught. There will be more opportunities over the course of the year in which there will be enough water in the channel to carry it to the sea. The risk of schedule delay will also be less for a vessel that can complete the passage to deeper water more quickly, because it can take advantage of lower, shorter-duration high tides to reach the sea. Transport options that have the lowest draughts and the greatest speeds will thus present the least schedule risk, while those having the greatest draughts and lowest speeds will present the greatest risk. Those mixing a high speed and draught or a low speed and draught will be intermediate in risk. These four possible pairings of speed and draught correspond very closely to the transport options (including the fuelled, baseline option) we identified earlier:

- **Fuelled submarine.** Deep draught, in part from the requirement to trim down at the stern to allow the propulsion system to work effectively, and high speed because it is under its own power.

**Table 2.1**
**Navigational Risk Assessment of Transportation Options**

| Clean Tow | Floating Transport Cradle | Heavy-Lift Ship |
|---|---|---|
| **High Risk** | **Medium Risk** | **Low Risk** |
| Towed submarine potentially impossible to manoeuvre in strong out-going tidal stream in Walney Channel. Slow. Very limited safety margins. Difficult pilotage. Once committed, minimal flexibility even with staging point. | Tow would require skilful handling. Enhanced safety margins. Slow. Flexible. Grounding requirement could be included in design. | Enhanced safety margins. High speed. Flexible. Can ground in emergency. |

- **Clean tow.** Deep draught and slow speed.
- **Floating transport cradle.** Draught reduced by design to the most cost-effective depth and here considered as shallow, and slow speed.
- **Heavy-lift ship.** Draught variable by ballasting once loaded, here considered as shallow, and high speed.

As shown in Table 2.2, the options can be arrayed along a gradient of decreasing schedule risk (right to left). The clean tow, with low speed and deep draught, is most vulnerable to delay, while the heavy-lift ship, with high speed and shallow draught, is the least so. The other two options fall in between. We judge the fuelled submarine to incur higher schedule risk than the floating transport cradle because the exit opportunities vary substantially over a relatively small range of draughts for a given speed. Additionally, small changes in draught in the 10-metre range comparable to the Astute imply large changes in exit speed. The floating transport cradle has flexibility to adjust the draught, thereby reducing schedule risk.

### Cost

Approximate costs for the transport options are listed in Table 2.3. The fuelled departure attracts no additional costs than currently planned, since we assumed that the current maintenance dredging would be the same for all cases. The BAE Systems shipyard at

**Table 2.2**
**Impact of Speed and Draught on Schedule Risk of Transportation Options**

| «—Decreasing Schedule Risk— | | | |
|---|---|---|---|
| High speed | Low speed | High speed | Low speed |
| Shallow draught | Shallow draught | Deep draught | Deep draught |
| Unfuelled | Unfuelled | Fuelled | Unfuelled |
| **Heavy-lift ship** | **Floating transport cradle** | **Submarine under its own power** | **Tow** |

**Table 2.3**
**Costs of Different Transportation Options**

| | Fuelled | Unfuelled | | |
| | | Clean Tow | Floating Transport Cradle | Heavy-Lift Ship with Cradle |
|---|---|---|---|---|
| Capital work | None | £20 million (staging point including capital dredging) | £9 million | £9 million |
| Tugs/heavy-lift ship | Not applicable | £0.5 million/sub | £0.5 million/sub | £1 million/sub |

Barrow-in-Furness is charging maintenance dredging solely to the Astute programme. If the site gets additional shipbuilding contracts in the future, these costs could be shared across multiple programmes. However, we do not take that into account here.

The capital work for the clean tow option is for the construction of the staging point. This includes the laying of mooring buoys, the capital dredging required for the staging point, and additional dredging of Walney Channel to make the turns feasible during towing. BAE Systems provided an assessment of cost for construction of the floating transport cradle, and this was included in the heavy-lift ship option assessment to ease loading and unloading of the submarine. In this way we did not need to assess the cost of building a very deep submergence heavy-lift ship for this purpose alone.

The last row of the table provides rental costs for ship hire—tugs in the case of the first two unfuelled options, the heavy-lift ship for the third. We based these costs on first-order estimates provided to us by marine salvage experts.

## Choice of Option

The clean tow option was eliminated because of its high navigational and schedule risks and high cost. The heavy-lift ship has a lower navigational and schedule risk than the floating transport cradle for a £0.5 million increase in recurring cost per submarine. A decision would have to be made as to whether the decrement in risk is worth the cost, but it should be kept in mind that the beam restrictions imposed by some of the gates in the lock and dock basin system could necessitate a purpose-built vessel that would add further costs. Also, the heavy-lift ship option might require a floating transport cradle anyway to aid in loading (and we assume that in our cost analysis). Using the floating transport cradle alone is thus a little less expensive and complex than using the heavy-lift ship. The floating transport cradle was therefore assessed as having the best balance of cost and navigational and schedule risks. We thus chose the floating transport cradle as the transportation option to assume in the remainder of our analysis. In summing costs across the dimensions covered in other chapters, those of the floating transport cradle are the costs associated with transportation.

## Regulatory Environment and Exit Plan Flexibility

The combination of international regulations for safe navigation passed in 2002, and the changed approach to nuclear regulation could affect exit plans for unfuelled or fuelled submarines. During our work, we sensed some uncertainty as to how the various transportation options and the exit safety requirements might be treated in an environment governed by current regulatory thinking. For these reasons, it is not prudent to assume that the Vanguard class exit plan could be adopted with little modification for Astute. As suggested by Figure 2.5, exit opportunities (and thus schedule delay) depend strongly on the eventual precise draught of the vessel and the precise time at which the boat leaves the Barrow dock area. It would thus be beneficial if the concerned parties could cooperate to craft a flexible exit plan taking advantage of modern technology to maximise exit opportunities while satisfying regulatory concerns. The following paragraphs address selected aspects of exit planning for which it appears possible to increase the flexibility for future submarine departures, fuelled or unfuelled, from Barrow-in-Furness. (To avoid raising the classification of this report, the following examples are deliberately not specific.)

Earlier exit plans used a set channel design that specified minimum required depths at a number of stages. Capital dredging was undertaken to establish the channel profile, and maintenance dredging has been undertaken in an attempt to keep the channel to this profile. However, some siltation to varying degrees along the channel is unavoidable between dredgings. The exit plans thus applied siltation margins, along with a survey error, that reduced the assumed depth from that specified in the dredging plan. Modern lightweight synthetic aperture sonars could be used much more closely to the anticipated departure date for Astute to confirm that Walney Channel is at least as deep as the defined profile.[6] The exit plans should then no longer need siltation and survey margins and could be written to allow for the actual sonar-determined depths to be used.

Exit flexibility could also be increased by improvements in the ability to dynamically predict the increases in allowable safety margin for the given weather and sea conditions, or even those forecast. Earlier plans required an almost doubling of the safety margin for the latter stages of the exit as the submarine approached the open area of the Walney Channel. This was to allow for maximum swell conditions and keep the pitching and rolling vessel within safety margins. Today, the swell encountered on departure can be predicted with more certainty, so that prediction can be relied on in place of the maximum. Furthermore, Astute movement characteristics, such as squat, pitch, and roll, are likely to be different to those of previous classes of submarine.

The mitigation of restrictions imposed by minimum visibility requirements will require more extensive examination than we can undertake here. The aim would be to take advantage of the much more precise and accurate positioning systems available today to reduce the impact of poor visibility. To win the confidence of those

---

[6] For example, Associated British Ports Barrow has a Reson Seabat 8101 swathe echo sounder that gives 100 percent seabed coverage that could be used for a full channel survey shortly prior to any exit.

navigating the Walney Channel, any visibility-related exit plan revisions would need to inform every aspect of the preparation of the channel, including dredging, positioning of markers, and surveying.

All three of the examples discussed above will benefit from the improved weather forecasting that is now available within the United Kingdom. Thus, as a submarine's planned departure date approaches, a flexible exit plan would allow refinement of the nominal plan for departure, for example, by opening a previously closed tide.

At this point, however, our analysis can only be suggestive. The safety conditions for the Astute have to be defined before it will be possible to rigorously describe schedule risk and allow a more direct comparison between the fuelled and unfuelled options.

BAE Systems is currently working with DELFT, a subcontractor, to revalidate the channel design used for the Vanguard exit plan. BAE Systems and DELFT are also updating sailing and hydrodynamic parameters, factoring in modern technology and towing techniques and plan to have the appropriate safety cases to the regulators by 2007.

## Conclusions and Recommendations

The transportation of an unfuelled submarine from Barrow-in-Furness through the Irish Sea to Devonport is feasible. Our analysis of transportation issues led to the following conclusions:

- In the transport route from Barrow-in-Furness to Devonport, the Barrow Exit has the most challenges for a fuelled or unfuelled submarine.
- A floating transport cradle and heavy-lift ship options both offer exit solutions. Given the information available, the cradle option was chosen as having the best balance of cost and navigational and schedule risks in our relative assessment, and we carry that option into the subsequent chapters.
- It is our view that the previous plan for the Vanguard class should not be replicated. Once the limits and conditions for safe

operation are defined for Astute, a flexible exit plan would greatly benefit the Astute programme by minimising schedule risk in the current regulatory environment, provided the new plan can be justified.

Exit from Barrow is a potential source of Astute programme schedule delay, which applies to all options, whether the submarine is fuelled or not, but there is an opportunity to significantly reduce schedule risk. First, however, BAE Systems (or the MOD) must promptly identify the safety conditions that will apply to an exit of Astute from Barrow-in-Furness. We say *promptly* because it may take some time to formulate an exit plan and secure agreement on it from all interested parties.

# CHAPTER THREE

# Facilities at DML[1]

The Devonport Dockyard operated by DML is the United Kingdom's only facility capable of refuelling nuclear submarines. Additionally, it refits and decommissions them. Thus, at some point in their existence, all UK submarines dock at Devonport. The operational fleet now includes four Swiftsure- and seven Trafalgar-class SSNs, and four Vanguard SSBNs. The Swiftsure and Vanguard classes operate from the Royal Naval Base at Faslane on the Clyde, whereas the Trafalgar class is based at the Royal Naval Base in Devonport. DML also provides operational support to all ships operating out of that base, as well as to visiting Royal Navy ships and those of other countries.

The DML dockyard is thus a busy one. If it is to fuel newly built submarines, the availability of suitable docks must be determined. Although the intention in consolidating fuelling and refuelling activities at DML would be to save money, there may be costs involved in doing so, and those should be estimated. Schedule risks should also be assessed. Those are the tasks we take up in this chapter, following an overview of maintenance activities at DML and a brief summary of the facilities on site.

---

[1] Numerous details about the facilities at DML and their usage have been omitted from this chapter to allow a general distribution of the report.

## Submarine Maintenance Activities

Numerous maintenance activities keep DML's facilities busy. Routine maintenance work ensuring the operational safety of a nuclear submarine is performed periodically during a submarine's in-service life. Some of these maintenance actions involve the use of a dry dock, especially for any mid-life refuellings. Also, at the end of its operational life, a submarine is subjected to a decommissioning, defuel, and lay-up period (DDLP) requiring a dock at DML.

The DML dockyard is currently busy with maintenance activities related to both SSNs and SSBNs. These demands on DML's facilities highlight the need to investigate their availability for fuelling of the Astutes, which has not been planned into the DML docks' schedules.

## Overview of Facilities

At DML, the NII-licensed/NNRP-authorised site for submarine maintenance and support is focused on facilities and docks around 5 Basin. Of the docks in 5 Basin, only those numbered 9, 10, 14, and 15 are currently being used for various submarine maintenance activities.

Since 1997, Devonport has received substantial government investment in successive phases of a project (D154) to modify facilities to refit and refuel Vanguard-class submarines and to provide modernised facilities for the refitting, refuelling, and defuelling of attack submarines or SSNs. Through this effort, 9, 14, and 15 docks have been upgraded and strengthened to modern seismic standards.[2] Our analysis of initial fuelling Astutes at DML focused on these docks.

---

[2] Devonport Management Limited, *Options for the Initial Fuelling of ASTUTE Class Submarines, Issue 1*, report provided to RAND on April 28, 2004.

## Availability of Docks to Fuel Astute-Class Submarines

We first looked at those dry docks that were capable of handling an Astute-class submarine. For each such dock, we considered the current schedule of in-service submarine work and any other actions that might eliminate the dock from use for a period of time. For some maintenance activities, a submarine may stay in a dry dock to do work that could be done at pier side. We also looked at moving submarines from a dry dock as early as possible and finishing any remaining work at the pier. As a result of the analysis, it seems unlikely that fuelling of Astute 1 could begin before mid-2008, eight months after the boat is scheduled for launch at Barrow. Beyond the first Astute, the remaining submarines can be accommodated, since the future docking periods need not take place precisely when indicated. As a result, the delays for future submarines would be minimal.[3]

## Schedule Risks for Refuelling Cases vs. the Baseline

Fuelling Astutes at DML would directly affect the programme schedule. This is true only of case 1, however, which assumes all submarines are fuelled at DML beginning with Astute 1. This case would be affected because of the unavailability of the dock to fuel the first of class on time. Case 2 assumes that Astute 1 would be fuelled at Barrow and that Astutes 2 and beyond would be fuelled at DML. Those boats could be accommodated, given some variability in the out-year docking schedule. The same is true of case 3, which assumes fuelling Astutes 4 and beyond at DML.

Table 3.1 summarises the schedule risks for all the three cases. The first case has the highest risk because of the eight-month sched-

---

[3] Further delays are possible if boats in service require unscheduled maintenance. We have not attempted to model the frequency, duration, or implications of such delays. We note, however, that fuelling an Astute will require only four months in a dock. Thus, assuming an 18-month Astute production drumbeat, there will be an Astute in a dock for only four of every 18 months.

**Table 3.1**
**Impact of 14 Dock's Availability on Schedule Risk for All Three Cases**

|  | Transportation | Facilities | Workload Impact | Nuclear Regulatory Issues | Contractual Issues |
|---|---|---|---|---|---|
| Baseline |  | No delay |  |  |  |
| Case 1 | BAE Systems action item | 8-month delay for Astute 1 |  |  |  |
| Case 2 |  | Minimal delay |  |  |  |
| Case 3 |  | Minimal delay |  |  |  |

NOTE: See pages 5 and 6 for the definition of each case.

ule delay for fuelling Astute 1. Cases 2 and 3 have minimal schedule risk relative to the baseline. (Avoidance of schedule delays associated with transportation is given as an action item for BAE Systems, consistent with the recommendations in Chapter Two.)

## Investment in Facilities

Besides the impact on schedule, some investment would be required to facilitate initial fuelling. Our discussions with DML and the ASM-IPT indicated that the equipment used to fuel the submarine at Barrow would be suitable for use at DML and could be provided as government-furnished equipment to DML. If this proves correct, the only cost incurred in this case would be related to transporting this equipment, which should cost roughly £0.25 million.[4] Additionally, the cost of renting a mobile crane amounts to approximately £50,000 per submarine.

As we proceed through the analysis, these costs will partly offset savings to be estimated in later chapters. We begin building a savings table here (Table 3.2) by entering the costs estimated so far as negative savings (in parentheses). Overall, the one-time facilities-related cost and the facilities-related cost per submarine to fuel Astute-class

---

[4] Based on our discussions with the CSALMO.

Table 3.2
**Facilities-Related Savings from Fuelling at DML for All Three Cases (in £ millions)**

|          | Transportation | **Facilities**      | Workload Impact | Nuclear Regulatory Issues | Contractual Issues |
|----------|----------------|---------------------|-----------------|---------------------------|--------------------|
| Baseline | None           | None                |                 |                           |                    |
| Case 1   | (9) + (0.5)/sub | (0.25) + (0.05)/ sub |                 |                           |                    |
| Case 2   | (9) + (0.5)/sub | (0.25) + (0.05)/ sub |                 |                           |                    |
| Case 3   | (9) + (0.5)/sub | (0.25) + (0.05)/ sub |                 |                           |                    |

NOTE: See pages 5 and 6 for the definition of each case.

submarines at DML are the same for all three cases. Total facilities-related costs would thus vary across cases only because the number of submarines fuelled at DML does. Case 2 involves fuelling two more boats at DML than in case 3 and would thus cost an extra £100,000, and case 1, which involves fuelling three more boats at DML, would cost an extra £150,000.

In summary, while the cost to fuel is minimal for all three cases, case 1 has the highest schedule risk with respect to fuelling of Astute 1 on time at DML. There is minimal risk of schedule delay with cases 2 and 3, corresponding to fuelling Astutes 2 and beyond at DML and Astutes 4 and beyond at DML, respectively.

# CHAPTER FOUR

# Workload Impact

What are the implications of fuelling new boats at DML for the allocation of work between BAE Systems and DML? To answer this question, we begin by comparing the total time required for delivering an operational submarine from Barrow to the MOD with the alternative of fuelling it at DML and completing the remaining work there. We then estimate the additional oversight effort required of DML personnel at Barrow and vice versa to ensure a smooth transition between the two yards. The cost and schedule risks are analysed across the three cases and compared against the baseline of completing all the work at Barrow. As prelude to the discussion of workload issues, we provide a brief overview of the submarine manufacturing process.

## The Submarine Manufacturing Process

In a typical submarine manufacturing process at Barrow, the hull units (or cylinders) are delivered to the Devonshire Dock Hall to be loaded with equipment in the final build line. The main machinery package is built and tested elsewhere on site prior to insertion into the open end of the aft hull units in the final build line. A number of process paths proceed in parallel; among them are paths for the nuclear steam-raising plant (NSRP), weapons, and ship systems paths. These paths merge during the final assembly and integration period before launch.

## Fuelling at Barrow

Fuelling is an integral part of the process by which Astute-class submarines are built. Once all the work is completed on the hull and the casing, the submarine is checked for watertight integrity. It is then moved out of the construction site (the Devonshire Dock Hall in this case) and launched into the dock using a synchrolift. Here, the submarine undergoes water testing of all systems (e.g., weapons handling, navigation, hydraulics) in parallel with postlaunch testing of the reactor and propulsion systems. Then, the boat must wait for a sufficient tide to exit the channel and begin sea trials. Based on calculations emanating from the analysis described in Chapter Two, a tidal window of one to three weeks is inferred.

## Schedule Risk of Fuelling at DML

To fuel at DML, the submarine has to be completed to a point at which it can be transported to Devonport. The prelaunch non-propulsion system work, which takes 15 weeks, still has to be completed prior to submarine launch and transport to DML and can no longer be carried out in parallel with fuelling. Once launched, the submarine would have to be tested[1] for a few days for its seaworthiness followed by waiting for the right tide over a period of one week. That is a shorter tidal window than that for the baseline because the submarine is assumed to have a floating transport cradle that would reduce draught, thereby increasing exit opportunities from Barrow as discussed in Chapter Two. Once out of Barrow, the submarine would complete the Irish Sea passage in one week before entering Devonport and 5 Basin from Plymouth Sound, which adds

---

[1] At this point, the submarine needs only to be subjected to an 'incline' test to ensure that it does not roll over during the time it is in water while being transported. This test is expected to take a couple of days. Extensive 'trim and incline' tests prior to sea trials need to be conducted after the submarine is fuelled, where a 'trim' test is required for submerging the submarine. These tests typically take about four days.

another week. Other minor activities connected with the move bring the net result to about 20 weeks, or five months, more to complete the submarine if fuelling were to take place at DML instead of at Barrow.

So far in the analysis, we have considered the fuelling process described earlier for Astute 1 at Barrow as the baseline. For subsequent boats, BAE Systems is planning to take a different approach.[2] BAE Systems claims that this process reorganisation would reduce the schedule by one month and save approximately 160,000 labour hours. The plan is to fully incorporate this process by Astute 3. This plan is impractical to undertake in the event of fuelling at DML. Thus, for Astute 3 and beyond, the savings in schedule and labour hours from the new plan should be viewed as an additional penalty for fuelling there.

We do not quantify any efficiency losses or their cost and schedule effects in cases 2 and 3. We would expect some such losses relative to the baseline and case 1 because fuelling and performing post-fuelling tasks on fewer boats at a given location should result in reduced opportunities for learning at that location.

In summary, completing Astute 1 and 2 would be delayed five months if they are fuelled at DML instead of at Barrow. Subsequent boats would also be delayed five months, plus an extra month relative to the new Barrow fuelling plan. Thus, for all three DML fuelling cases, the schedule delay would amount to six months for Astute 3 and beyond (five months for Astute 1 and 2). Relative to the baseline, all DML cases are thus denoted as having high schedule risk, as shown in Table 4.1.

---

[2] This should not be confused with a recent proposal by BAE Systems to reduce nuclear hazards by locking the control rods in place after fuelling and possibly conducting power range testing at Faslane (see Chapter Seven).

**Table 4.1**
**Workload-Related Schedule Risk for All Three DML Fuelling Cases Compared Against the Baseline**

| | Transportation | Facilities | Workload Impact | Nuclear Regulatory Issues | Contractual Issues |
|---|---|---|---|---|---|
| Baseline | | No delay | No delay | | |
| Case 1 | BAE Systems action item | 8-month delay for Astute 1 | Up to 6-month delay | | |
| Case 2 | | Minimal delay | Up to 6-month delay | | |
| Case 3 | | Minimal delay | Up to 6-month delay | | |

NOTE: See pages 5 and 6 for the definition of each case.

## Cost of Fuelling at DML

The workload cost or savings from fuelling at DML depends on differences in direct labour hours and wage rates at the two sites. The total manufacturing labour hours in the baseline for completing all the work at Barrow should be comparable to splitting the effort between the two sites, as the same tasks must be accomplished. Direct wage rates at the two sites are comparable to each other. With respect to the overhead rate, at Barrow it accrues to the Defence Procurement Agency, and at DML to the Defence Logistics Organisation; either way, the MOD ultimately shoulders the cost. As a result, we concluded the total production labour costs of fuelling at DML to be comparable to the baseline of doing all the work at Barrow.

Such an operation would involve oversight costs, which are not part of the manufacturing costs. DML personnel should monitor construction at Barrow before receiving the boat for further work. BAE Systems personnel should be available at DML for any input that could be useful to the continuation of the production process that was started at Barrow. Approximately eight skilled BAE professionals would be required from two elements: the dockside test organisation responsible for testing the ship systems and weapons, and the reactor test group for testing the NSRP. This is estimated to cost approximately £0.5 million per submarine. Likewise, a staff of five DML personnel would need to be stationed at Barrow to oversee the manufacturing process three years immediately prior to sailing.[3] This is estimated to cost £1.7 million per submarine; the total is thus £2.2 million per fuelling operation at DML. Besides these oversight costs, savings of the new fuelling process on Astute 3 and beyond that cannot be realised at DML are added on as a penalty for fuelling there. Assuming an average direct wage rate of £10 per hour, the BAE Systems estimate of a 160,000-man-hour savings translates into an additional cost to the DML cases of approximately £1.6 million per submarine for Astute 3 and beyond. Rolls-Royce currently provides engineering support at Barrow for the NSRP. This is assumed to be transferable to DML.

As indicated above, if the core is loaded at DML instead of at Barrow, the boat will spend an extra 15 weeks in the shipyard—specifically, at DML—because the core load cannot be accomplished in parallel with other activities. The longer a ship is in a yard, the more berthing and production support costs (riggers, security, temporary power, etc.) it incurs. According to DML, however, no extra docking or berthing costs will accrue for tying up a submarine at its facility. These costs are considered part of the overhead to maintain nuclear infrastructure, which is paid by the Warship Support Agency through the Defence Logistics Organisation chain, regardless of the number of submarines using the facility. According to DML, the only extra expense would be for electricity to power the ship systems over an extended period of time. Considering the delay of 15 weeks at DML, this cost was estimated to be £0.1 million per submarine being fuelled there.

The total cost per submarine for the workload-related issues thus amounts to £3.9 million per submarine for the three cases and is entered into the savings table (Table 4.2) as negative numbers.[4] This

---

[3] Henry Buchanan and Dan Wiper, *A2B Fuelling Study*, BAE Systems, ASTUTE Class Project, Reference 01/00/41000000/RP/8098026(5), Astute Issue 3, 7 June 2002.

[4] Efficiency-related savings due to learning are not considered here, as mentioned earlier.

recurring cost includes DML electricity charges of £0.1 million, £1.6 million considered as a penalty for not taking advantage of the new process at Barrow and fuelling at DML, and £2.2 million for the exchange of oversight teams. It is noteworthy that these costs rely on estimates provided by BAE Systems.

**Table 4.2**
**Workload Impact on Savings from Fuelling at DML (in £ millions)**

|  | Transportation | Facilities | Workload Impact | Nuclear Regulatory Issues | Contractual Issues |
|---|---|---|---|---|---|
| Baseline | None | None | None |  |  |
| Case 1 | (9) + (0.5)/sub | (0.25) + (0.05)/sub | (3.9)/sub |  |  |
| Case 2 | (9) + (0.5)/sub | (0.25) + (0.05)/sub | (3.9)/sub |  |  |
| Case 3 | (9) + (0.5)/sub | (0.25) + (0.05)/sub | (3.9)/sub |  |  |

NOTE: See pages 5 and 6 for the definition of each case.

# CHAPTER FIVE

# Nuclear Regulatory Issues

Nuclear submarine manufacture and maintenance in the United Kingdom is subject to regulation by the MOD's Naval Nuclear Regulatory Panel and the Health and Safety Executive's Nuclear Installations Inspectorate. Although the nuclear regulations have not changed since the last of the Vanguard class, the manner in which they are implemented has changed, resulting in additional costs of compliance for manufacturing the Astute-class submarines at Barrow. This site is currently in the process of preparing safety cases for initial fuelling and critical operations for the new class after a lapse of almost a decade since the last Vanguard-class nuclear submarine was fuelled and tested there. Barrow had originally projected a cost of approximately £20 million for this effort, which was based on the regulatory environment in which the Vanguard class was manufactured. As a result of changes in the environment since then, anticipated site licensing costs have increased by an additional £100 million over the original proposal.

This chapter begins with an overview of the regulatory environment by addressing specifically what has changed since the Vanguard class. We then take up the implications of the regulatory environment for fuelling at Barrow and at DML. For Barrow, we are interested primarily in how much of the increased licensing and authorisation costs could be saved by moving fuelling to DML, in the costs of decommissioning the site, and in subsequent annual savings of regulatory expenses. At DML, we are interested in what the contractor will have to do to comply with regulations to facilitate fuelling new boats there. The chapter concludes with the ramifications of these assessments for cost and schedule risk accruing to each of the three DML fuelling cases and the baseline.

Unfortunately, substantial uncertainty must be attached to our cost estimates. In particular, our estimates may be conservative because the regulatory environment is nonprescriptive. That is, work needed to gain regulatory approval is not specified or bounded by the regulators. Moreover, the MOD's contracts cover the cost of efforts needed to satisfy the regulators. This combination encourages very conservative safety cases and mitigating measures to ensure that the product clears the bar for approval. However, the uncertainty does not importantly limit our conclusions.

## The Nuclear Regulatory Environment

Under the Nuclear Installations Act of 1965, no site may be used to install or operate any nuclear facility unless the Health and Safety Executive (HSE) has granted a license. The BAE Systems and DML shipyards in Barrow and Plymouth, respectively, are licensed and regulated by the NII, which resides within the HSE.

The 1965 act contains an exemption for a 'nuclear reactor comprised in a means of transport' (i.e., a submarine reactor plant). However, the Secretary of State for Defence has a policy that, where the MOD is exempt from regulation, it will, so far as is reasonably practicable, have policies at least as good as the regulations. In this vein, the MOD appointed the Chairman, Naval Nuclear Regulatory Panel (CNNRP) as its regulator with responsibility for establishing and maintaining standards and arrangements for the Naval Nuclear Propulsion Programme.[1] The CNNRP's main area of concern is the submarine, whereas the NII's main focus is on safety of operations

---

[1] Nuclear Installations Inspectorate, and Chairman, Naval Nuclear Regulatory Panel, letter of understanding, 6 March 2003.

within the licensed site.[2] Nevertheless, there is some dual regulation in cases where nuclear propulsion plant work is performed at a privately owned site, as discussed below.

Both organisations operate under similar regulatory frameworks with conditions ('License Conditions' to the NII, 'Authorisation Conditions' to the CNNRP) defining areas of nuclear safety. NII nuclear site licensees (and/or CNNRP authorisees) must comply with these conditions, which range from arrangements for ensuring plant safety and controlling operations to management issues such as supervision and training of staff.

Both regulators operate in a nonprescriptive regime, which places reliance on self-regulation by the licensee or authorisee. In this approach, neither regulator prescribes safety standards that a licensee or authorisee is expected to follow. Instead, both organisations use similar Principles (the NII's Safety Assessment Principles; the CNNRP's Safety Principles and Safety Criteria)[3] to assess the safety cases required by the license and authorisation conditions. Specifically, Condition 14 requires licensees and authorisees to 'make and implement adequate arrangements for the production and assessment of safety cases consisting of documentation to justify safety during the design, construction, manufacture, commissioning, operation and decommissioning phases of the installation'.

In 1996, the NII and the MOD agreed to change their approach to dual regulation. The MOD agreed to provide the HSE licensee with data on the design of the submarine nuclear reactor. The intent was not for the NII to look into the reactor design but instead for it to gain understanding of the reactor's safety-related matters and how these interacted with shore-based facilities.[4] Today, the NII and the CNNRP regulate nuclear submarines jointly. They describe their respective roles and responsibilities in a 2003 letter of understanding in which they agreed to share information, endeavour

---

[2] Refer to UK National Audit Office (2002), p. 13.

[3] CNNRP email, 28 January 2004.

[4] UK National Audit Office (2002), p. 13.

to jointly determine and agree on any action, and take all reasonable steps in deciding which organisation should take action.[5]

Although the NII's Principles were last revised in 1992 and the CNNRP issued its Principles in 1994, both DML and BAE Systems have asserted that regulatory changes in the 1990s greatly increased costs. In DML's opinion, the 1996 Ministry of Defence/Health and Safety Executive Agreement fundamentally altered the means of dealing with the interface between the submarine's nuclear reactor and the dockyard's facilities, requiring substantial extra work and cost for DML on the D154 project in upgrade of submarine facilities at its site.[6] The National Audit Office, in its review of this project,[7] reported that the practical challenges and subsequent cost effects of how the nuclear regulation regime would affect this project were not fully appreciated by any party. The imperative to meet milestones to support the submarine refit programme also contributed to cost increases; the facilities' design evolved to take account of the regulators' observations, requiring additional construction work. Similarly, BAE Systems asserts that nuclear regulatory costs have increased as a result of the adoption and implementation of safety management arrangements needed to produce and comply with the post-1996 suite of safety documentation. BAE Systems describes the current safety cases as the first on the Barrow site for which a comprehensive Site Safety Justification has been required for all operations carried out to build, commission, and test nuclear-powered submarines.

The change to joint regulation and NII access to reactor plant design data made licensees address all hazards in their safety cases, including the interrelationship of site and reactor plant hazards. Addressing this interrelationship adds complexity, with corresponding effort and cost, in preparing and implementing post-1996 safety

---

[5] Nuclear Installations Inspectorate, and Chairman, Naval Nuclear Regulatory Panel, letter of understanding, 6 March 2003.

[6] UK National Audit Office (2002), p. 24.

[7] UK National Audit Office (2002).

cases, even though the Principles by which the two regulators assess safety cases have not changed.

Additionally, the nuclear regulatory cost for the Astute contract increased because of the new requirement for CNNRP authorisation of the Barrow site, which was provided in November 2004.

## Issues at Barrow

To comply with the NII/CNNRP licensing and authorising conditions applicable during construction, testing, and delivery of the first three Astute submarines, the MOD will incur costs possibly in excess of £100 million. This amount is based on regulatory compliance costs incorporated in the original contract and additional costs now anticipated, including

- authorisation of the Barrow site
- self-regulation by BAE Systems
- production and compliance with a modern style suite of safety documentation.

It should be noted that estimates of nuclear regulatory compliance costs are based on today's conditions and regulations. Further changes in regulations or regulatory enforcement policies and practices could substantially alter these estimates. History suggests, though, that nuclear regulatory compliance costs are more likely to increase than decrease in the future.

Nuclear regulatory compliance costs are both recurring and nonrecurring. Examples of recurring costs include training, maintenance of safety committees, and periodic safety reviews. Nonrecurring costs principally involve preparation of safety cases to satisfy NII Condition 14 and, for any operation that may affect safety, '...to demonstrate the safety of that operation and to identify the conditions and limits necessary in the interests of safety' (NII Condition 23).

Transfer of fuelling operations to DML could allow BAE Systems to end its NII license at Barrow and avoid incurrence of some further nuclear regulatory compliance costs at Barrow. However, even without a license, certain nuclear regulatory compliance costs would remain at Barrow. Maintenance of an NSRP fabrication quality control system operated by suitably qualified and experienced personnel is an example of a nuclear regulatory cost that has to remain at Barrow even if fuelling and initial criticality occur elsewhere.

Figure 5.1 shows the estimated spend-out profile for the £71 million of remaining Barrow nuclear regulatory compliance costs for the first three Astute-class submarines.[8] For example, it is anticipated that £30 million of compliance costs will be spent by 2006, leaving £40 million. As such, the graph provides an estimate of gross savings that may still be available if at any time a decision is made to fuel Astute submarines elsewhere. The savings decline rapidly as the Barrow yard completes safety cases and incurs recurring costs. A prompt decision maximises potential savings. Additionally, once fuel arrives at Barrow for the first submarine, all nuclear activity for that boat up through power range testing (PRT) will have to occur at Barrow, meaning that all safety cases would have to be completed. In this case, savings drop to those available when the first ship leaves Barrow in 2008, even if a decision to move subsequent ships is made before 2008.

It is important to note that the graph shows only the *savings* available at Barrow; it excludes any licensing costs that must be incurred at Barrow for nuclear-related activities whether or not fuelling is undertaken there. It also excludes the costs of the safety cases for the boats, which must be prepared regardless of where they are fuelled. Extra costs required to achieve license and authorisation conditions at DML should be subtracted from the savings shown in Figure 5.1. However, we have no estimate of such costs but have been assured that, because DML is already licensed for the more challeng-

---

[8] The original contract was for approximately £20 million. The additional £100 million now anticipated brings the total amount to around £120 million. With £30 million required to maintain manufacturing quality assurance standards at Barrow—even if no fuel arrives there—and £19 million already spent, about £71 million would remain unspent towards nuclear regulatory compliance for the first three boats.

ing task of handling spent fuel, they will be much smaller than the costs at BAE Systems (shown as savings in the figure).

Any decision to stop fuelling at the Barrow yard and terminate the site license will require the site to be decommissioned in accordance with NII Condition 35. BAE Systems' present estimate is that this effort would cost £15 million. Decommissioning expenses would reduce the potential savings available to the government from the Astute contract if new boats were fuelled at DML.

Not fuelling at Barrow will also result in savings in nuclear overhead and recurring regulatory expenses beyond Astute 3. Our discussions with BAE Systems personnel indicate the overhead savings to be minimal, on the order of £0.25 million per year. This is because most of the costs are charged directly to the Astute contract. Moreover, radiographic equipment that is currently part of the overhead would still be required to conduct inspection and testing to

**Figure 5.1**
**Available Savings Associated with Site Licensing Reduce Annually**

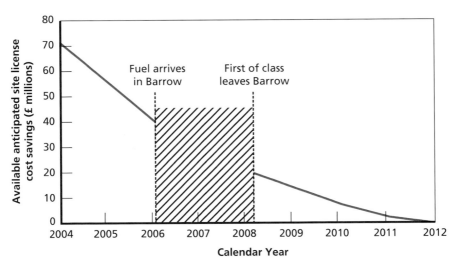

maintain manufacturing standards. As for the directly charged recurring regulatory expenses, a significant amount of quality assurance work would still be required even if no nuclear fuel were handled at Barrow. BAE Systems estimates that the currently projected recurring regulatory expenses to maintain the site license would fall from £10 million per year to £3 million per year.

## DML Issues

DML also has an NII license for its Plymouth site, where it presently refuels Vanguard- and Trafalgar-class submarines and defuels inactivated boats. The NSRP used in the Astute class is very similar to that used in Vanguard, so it is likely that DML's current Vanguard refuelling and related safety cases could be modified to apply to Astute fuelling. However, fuelling Astute at the DML licensed site presents some challenging regulatory issues:

- The NII is highly unlikely to approve Astute 1 fuelling in 14 Dock until the 80-tonne crane is removed and other submarine refuelling facilities are renovated. This would delay Astute 1 delivery, as noted in Chapter Three.
- As the licensee handling the fuel and taking the NSRP critical for the first time—DML, not BAE Systems—would have to satisfy the NII that the NSRP fabrication meets required standards. Having no involvement with the Astute contract to date puts DML at a great disadvantage, especially for the first Astute.
- The NII will require DML to demonstrate that it is in control of fuelling and PRT at its site, which will be difficult for DML to do with BAE Systems as the Astute's prime contractor and design authority.
- The NII could require DML to obtain a new license for fuelling Astutes, which would delay the process and require public involvement.

## Cost and Schedule Risk

Table 5.1 summarises the net potential regulatory savings from eliminating BAE Systems Barrow site license and performing all fuelling and PRT at DML. The savings shown are the gross savings from Figure 5.1 reduced by the estimated decommissioning costs at Barrow and an estimated £0.5 million cost for DML to prepare Astute fuelling and related safety cases. This estimate by DML is possibly conservative, but it is difficult to devise a more reliable one because the scope of the safety case is undefined without having a detailed plan.

Case 1, in which all Astutes are fuelled at DML, presents the greatest potential savings, depending on when a decision to transfer fuelling occurs. If it takes place immediately, the full £71 million gross from Figure 5.1 is available, minus £15.5 million, as mentioned above, for a net of £55.5 million. If a decision does not occur until 2006, just before fuel would arrive at Barrow, the net available drops to £24.9 million. In this case, as with the others, the annual £7.3 million in Barrow license maintenance and nuclear overhead savings are realised as soon as the site is decommissioned. In case 2, in which Astute 1 is fuelled and delivered from Barrow, only the £20 million potential savings left after fuel delivery is available from the gross savings in Figure 5.1, minus Barrow decommissioning costs and DML safety case costs. The latter costs are also incurred in case 3, in

### Table 5.1
**Nuclear Regulatory Savings for All Cases (in £ millions)**

|  | Transportation | Facilities | Workload Impact | Nuclear Regulatory Issues | Contractual Issues |
|---|---|---|---|---|---|
| Baseline | None | None | None | None |  |
| Case 1 | (9) + (0.5)/sub | (0.25) + (0.05)/sub | (3.9)/sub | 24.9 to 55.5 + 7.3/year |  |
| Case 2 | (9) + (0.5)/sub | (0.25) + (0.05)/sub | (3.9)/sub | 5.5 + 7.3/year |  |
| Case 3 | (9) + (0.5)/sub | (0.25) + (0.05)/sub | (3.9)/sub | (15.5) + 7.3/year |  |

NOTE: See pages 5 and 6 for the definition of each case.

which Astutes 4 and beyond are fuelled at DML, and there are no offsetting savings from the present Astute 1–3 contract. Not shown here are potential savings in all DML refuelling cases from any un-anticipated future tightening of the nuclear regulatory environment.

Although case 1 presents the greatest potential savings, it also presents the greatest schedule risk (see Table 5.2). Astute 1 will likely be delayed significantly because of concerns raised by regulators related to DML's lack of involvement in the programme to date, and the unavailability of its facilities for fuelling the submarine on time. Later Astutes would likely not suffer delays, since DML would have sufficient time to resolve regulatory and facility issues with the NII.

It is important to note that the baseline programme, in which all Astutes are fuelled and delivered from Barrow, will present moderate nuclear regulatory compliance schedule risk. Officials in several organisations related to the Astute effort opined that the regulatory work left on the Astute programme could well be the critical path to delivery of the first boat. As Figure 5.1 shows, there is a large amount of new regulatory effort for BAE Systems to accomplish in the next several years.

**Table 5.2**
**Schedule Risk Associated with Nuclear Regulatory Issues for All Cases**

|  | Transportation | Facilities | Workload Impact | Nuclear Regulatory Issues | Contractual Issues |
|---|---|---|---|---|---|
| Baseline |  | No delay | No delay | Uncertain scope |  |
| Case 1 | BAE Systems action item | 8-month delay for Astute 1 | Up to 6-month delay | NII concerns |  |
| Case 2 | | Minimal delay | Up to 6-month delay | No delay |  |
| Case 3 |  | Minimal delay | Up to 6-month delay | No delay |  |

NOTE: See pages 5 and 6 for the definition of each case.

# CHAPTER SIX

# Remaining Issues and Overall Summary

This chapter begins with an overview of contractual challenges that need to be resolved among key players, including the MOD, BAE Systems, DML, Rolls-Royce, the NNRP, and the NII, to facilitate the fuelling of Astutes at DML. Additionally, we estimate the cost of having a contract at DML in the form of an additional fee required. We address the strategic implications of consolidating fuelling and refuelling at DML and the implications for public perception of MOD actions at the two sites by making such a move. The chapter concludes with an overall summary of cost and schedule risks across all the quantified categories to rank the three cases against the baseline.

## Contractual Challenges

The process of fuelling at DML cannot be undertaken without executable contract arrangements performed by wholly cooperative entities. BAE Systems obviously does not want to lose work and future capability. It wants to protect the programme and to deliver the submarine on time, or earlier if possible, and to reduce risk. DML, however, is interested in additional business, which would make it the sole nuclear licensed dockyard capable of handling nuclear fuel in the United Kingdom. These opposing incentives mean that the MOD would have to take the lead to get the two parties to agree on an executable contract. Rolls-Royce as the NSRP supplier and the nuclear regulators also need to be highly involved to promptly address critical issues related to a fuelling transfer.

Fuelling at DML entails transporting a partially completed submarine to the site and finishing the remaining work there while BAE Systems retains design authority and the prime contractor's responsibility for manufacturing and delivering an operational submarine to the MOD. This translates to BAE Systems providing assurance to the MOD that work being done at DML meets the contract requirements. Nuclear regulators, however, require DML to have and demonstrate complete control over fuelling and PRT at Devonport. Reconciling these divergent government requirements will be a significant contractual challenge.

The ASM-IPT indicated that it might be necessary to take delivery of a partially completed submarine from BAE Systems and provide it as government-furnished equipment to DML. This would relieve BAE Systems of its responsibility to the MOD and provide a plausible solution. However, the parties involved would need to clearly spell out the details of who is ultimately responsible for what and write them into an executable contract including contractor sea trials and post–sea trial work. This contract should take into consideration the complications with regards to the responsibility for existing and emergent defects at a later date.

These contractual issues can be challenging and require some time. The MOD will have to drive the process and provide appropriate incentives to ensure that the key players, including BAE Systems, DML, Rolls-Royce, and the regulators are actively involved in setting up an executable contract. Because there is a contract in place for the first three boats, resolving contractual issues between BAE Systems and DML will present the greatest challenge for cases 1 and 2. For case 3, we assume that contract issues can be worked out, possibly with separate contracts or partner alliances.

## Contract-Related Costs at DML

Settling the contractual issues will require labour beyond that needed for the baseline. Under the scenario in which BAE Systems retains full responsibility for delivering the completed submarine to the MOD and the fuelling effort is subcontracted to DML, we anticipate DML would charge a fee on the additional work it will be doing. This fee would be in addition to the profit that BAE Systems would charge the MOD for manufacturing and delivering a fully operational nuclear submarine. Based on the data provided by BAE Systems personnel at Barrow, we estimated the fuelling and post-fuelling effort prior to sea trials at approximately 215,000 hours[1] distributed across management, outfitting, support, structural, technical, and test and commissioning disciplines. Assuming a typical fee of 5 to 10 percent on the fully loaded rate at DML, we estimated the cost to the contract to be about £0.5 million per submarine, which is the same for all three cases as shown in Table 6.1.

**Table 6.1**
**Potential Contract-Related Savings from Fuelling at DML (in £ millions)**

|  | Transportation | Facilities | Workload Impact | Nuclear Regulatory Issues | Contractual Issues |
|---|---|---|---|---|---|
| Baseline | None | None | None | None | None |
| Case 1 | (9) + (0.5)/sub | (0.25) + (0.05)/sub | (3.9)/sub | 24.9 to 55.5 + 7.3/year | (0.5)/sub |
| Case 2 | (9) + (0.5)/sub | (0.25) + (0.05)/sub | (3.9)/sub | 5.5 + 7.3/year | (0.5)/sub |
| Case 3 | (9) + (0.5)/sub | (0.25) + (0.05)/sub | (3.9)/sub | (15.5) + 7.3/year | (0.5)/sub |

NOTE: See pages 5 and 6 for the definition of each case.

---

[1] To estimate this total, we used Astute 2 hours because they are likely to be more representative than those of the first of class. These hours represent the total labour hours in the last three quarters prior to sea trials as provided by BAE Systems. Their data were not sufficiently disaggregated to allow us to separate out any hours that might not have to be performed at DML.

## Schedule Risks Associated with Contracting

Addressing all the contractual issues in time to fuel Astute 1 at DML can be a significant challenge, especially since the contract price has been negotiated. To motivate BAE Systems to negotiate, the MOD would probably need to provide financial incentives in excess of existing fee incentives; the additional incentives could well supersede any savings that may be realised due to reduction in work scope at Barrow. The difficulty of achieving new contractual arrangements acceptable to all parties, including the regulators, means there is a risk of delay with cases 1 and 2, in which the Astutes covered by the current contract are fuelled at DML. The hurdles in the way of on-time completion for Astute 1 are the most significant, since DML has had no involvement to date. This leads to case 1 having the highest schedule risk and case 2 having medium risk with relatively more time to address contractual issues prior to fuelling Astute 2 at DML, compared with the baseline of no contractual risk as shown in Table 6.2. There should, however, be sufficient time available to address all the contractual issues for subsequent contracts. Hence, we have rated case 3 as having no contractual delays.

Table 6.2
**Schedule Risk Related to Contractual Issues for All Cases**

| | Transportation | Facilities | Workload Impact | Nuclear Regulatory Issues | Contractual Issues |
|---|---|---|---|---|---|
| Baseline | | No delay | No delay | Uncertain scope | No delay |
| Case 1 | BAE Systems action item | 8-month delay for Astute 1 | Up to 6-month delay | NII concerns | Significant delay |
| Case 2 | | Minimal delay | Up to 6-month delay | No delay | Some delay |
| Case 3 | | Minimal delay | Up to 6-month delay | No delay | No delay |

NOTE: See pages 5 and 6 for the definition of each case.

## Public Perception

Initial fuelling at DML would result in a loss of work from Barrow and an equivalent gain at Plymouth. BAE Systems is the largest supplier of high-paying jobs in the Barrow area, which has relatively few other employment opportunities to offer to the local population. A recent study by Furness Enterprise[2] addresses the negative impact of making employees redundant, as well as the cascading effect on the local economy. Besides these economic issues and the low public morale related to it, key skills in assembling, testing, and commissioning the NSRP could be irreversibly lost from the area. This would make it extremely difficult if not impossible for the Barrow site to deliver a fully operational nuclear submarine in the future. Such a move would force Barrow to bid jointly with DML on future nuclear submarine manufacturing contracts, since DML would be the only site licensed to fuel the boats.

DML at Plymouth is currently very busy with long overhaul periods (for refuelling) of the Vanguard class and the last of the Trafalgar class and with decommissioning of the Swiftsure- and Trafalgar-class boats. Additional nuclear work at DML could have some, albeit minimal, delay on the decommissioning schedule. This could result in an accumulation of nuclear submarines waiting to be decommissioned for a short period of time. Such an accumulation could be cause for concern among the people living in the densely populated Plymouth area and would need to be approved in detail by the regulators. To ease these concerns, DML would have to take a proactive role in educating the local population about the safety precautions undertaken. Overall, however, community reactions to the transfer of fuelling to Plymouth are likely to be more strongly negative in Barrow than in Plymouth because of the loss of jobs from the Barrow area, where economic opportunities are limited.

---

[2] Furness Enterprise, *UK Submarine Industrial Base: Employment Issues Associated with Nuclear Steam Raising Plant in New Submarines*, 8 July 2004.

## Overall Cost and Schedule Summary

Here, we combine cost and schedule effects across all issues considered in this report. Because the favourability of these two effects is not aligned across cases, the two sets of results must be weighed against each other.

### Costs and Savings

To combine the costs and savings associated with all the issues and compare the sums for the three cases with each other and the baseline, we conducted a net present value (NPV) analysis. Since all three cases involve costs related to submarine manufacturing in the future, this analysis is especially important because expenditures or savings at different times in the future would have different current values. We thus derive a net present value of future costs and savings for all three cases, expressed in current-day pounds. We used a discount rate of 3.5 percent as listed on Her Majesty's Treasury Web site for the year 2004. Recent statements made by the Defence Secretary led us to assume a total of eight Astutes for the analysis, with an 18-month production drumbeat.

Table 6.3 provides the results of the NPV analysis. As shown, case 1 has the highest savings, followed by very modest savings for case 2, and no savings for case 3.

**Table 6.3**
**Net Present Value of Savings from All Three Cases Compared Against the Baseline (£ millions)**

|          | NPV   |
|----------|-------|
| Baseline | None  |
| Case 1   | 24–57 |
| Case 2   | 9     |
| Case 3   | (2)   |

NOTE: See pages 5 and 6 for the definition of each case.

## Schedule

When we look across all the issues for schedule risk as shown in Table 6.4, we find that case 1 has the highest risk, which is primarily related to the delays with respect to Astute 1. Those amount to at least 14 months beyond the baseline (significant contractual delays might push delivery back further). Six months of that delay accrue to all boats in all cases. That is, the entire programme is set back six months; once the six-month delay accrues to the first boat to be fuelled at DML, the others will emerge at intervals equal to the original production drumbeat.

## Synthesis

In summary, while case 1 may have the highest savings, high risk of schedule delays makes this case undesirable. Case 2 has very modest savings with relatively lower schedule risk, and case 3 does not save any money. Because savings was the main reason for considering the consolidation of fuelling and refuelling at DML, case 3 is the least attractive alternative when compared with the baseline.

**Table 6.4**
**Summary of Schedule Risk for All Three Cases Compared Against the Baseline**

|  | Transportation | Facilities | Workload Impact | Nuclear Regulatory Issues | Contractual Issues |
|---|---|---|---|---|---|
| Baseline |  | No delay | No delay | Uncertain scope | No delay |
| Case 1 | BAE Systems action item | 8-month delay for Astute 1 | Up to 6-month delay | NII concerns | Significant delay |
| Case 2 |  | Minimal delay | Up to 6-month delay | No delay | Some delay |
| Case 3 |  | Minimal delay | Up to 6-month delay | No delay | No delay |

NOTE: See pages 5 and 6 for the definition of each case.

# Emerging Issues

Throughout the course of this project, new ideas have been generated, researched, analysed and, where appropriate, incorporated into the study. At the very end of the study, after field research was completed, BAE Systems informed RAND that it had devised a new approach to fuelling and testing Astute-class submarines at Barrow that could significantly reduce nuclear regulatory compliance costs. Although we could not do a thorough assessment of that proposal, this chapter provides some details of the new approach, conditional on the accuracy and robustness of the information provided by BAE Systems.

## The New Approach

The BAE Systems concept is to take measures during construction to significantly reduce or eliminate potential nuclear consequences of hazards at Barrow. After taking these measures, the submarine would be completed, launched, and tested. In this regard, BAE Systems offered two options:

- **Baseline option 1:** Perform all testing at Barrow and exit for sea trials.

- **Baseline option 2:** Transport the fuelled ship to Faslane (an NNRP-authorised site), perform final testing there, and exit for sea trials.

The effort to produce safety cases in which there are no significant nuclear consequences for the public should be much less than that required for safety cases with potential nuclear consequences. On this basis, BAE Systems estimates that £18 million in present value could be saved on the Astute 1–3 contract if testing is performed at Barrow and £50 million if testing is performed at Faslane. This latter estimate does not include transportation costs to Faslane, safety case preparation and implementation for the transit to Faslane, and the safety cases needed to perform initial testing at Faslane. BAE Systems also advises that all three submarines could be delivered within their contract delivery dates under this approach.

This new approach appears to be a promising way to reduce Astute-class nuclear regulatory compliance costs with no increased risk, and perhaps reduced risk, to the public. Additionally, the moderate schedule risk assessed for regulatory issues in the baseline case would be eliminated. The NII confirmed that the basic premise for the savings generated by this innovative approach is valid, namely that producing safety cases in which significant consequences for the public are lower should be less costly.

## Conditional Cost and Schedule Assessment

Unfortunately, we could not analyse in detail the feasibility, savings, or schedule consequences of this approach. Nevertheless, assuming the savings quoted by BAE Systems to be valid, we conducted an NPV analysis to compare the savings with the three DML fuelling cases and the original baseline already considered in the analysis (see Table 7.1).

As shown, baseline option 2 takes into account the transportation costs for eight Astute-class submarines, amounting to £12 mil-

**Table 7.1**
**Net Present Value Analysis of the**
**Two New Baseline Options Compared**
**Against All Three Cases and the**
**Original Baseline (£ millions)**

|  | NPV |
|---|---|
| Baseline Option 1 | 18[a] |
| Baseline Option 2 | 38[a] |
| Baseline | None |
| Case 1 | 24–57 |
| Case 2 | 9 |
| Case 3 | (2) |

[a] These estimates are conservative with respect to savings—that is, the savings are likely greater.

lion present value, that are not incurred in baseline option 1. The savings for both options could be greater than those shown, since there should be additional savings in recurring nuclear regulatory expenses at Barrow.

## Uncertainties in the Assessment

While these options show promise, important uncertainties remain. We could not assess

- the availability of berths, services, or testing equipment necessary for initial Astute PRT at Faslane
- the effort required to prepare a safety case for initial Astute testing at Faslane
- any potential limitations on the number of days of critical operations that can be performed at Faslane
- issues of ownership of the untested submarine
- the availability of suitably qualified and experienced personnel.

It appears feasible to transport a fuelled, unpowered Astute from Barrow-in-Furness to Faslane in the same way that an unfuelled submarine might be moved to DML. The safety case, however, is likely to be different and may have to involve the MOD and the NII, as well as other government departments and agencies such as the Department for Transport and the Maritime and Coastguard Agency. We cannot comment on any potential impact that the involvement of these agencies might have on the complexity of safety case and the resources required.

The estimating details behind BAE Systems projected savings are unknown to us. Moreover, near-term savings could be reduced in the Faslane option if the NII insists on decommissioning the Barrow site. It might do so because performance of testing at Faslane would mean that there would no longer be any need to perform low-level radioactive work at Barrow, which eliminates the need to retain contaminated facilities. (A license would still be required to handle the unspent fuel loaded into submarines.) Decommissioning costs, together with transportation costs already mentioned, could eat up much of BAE Systems' projected savings for the first three Astute-class submarines under the Faslane option. However, this approach could still be desirable because savings would accrue to future submarines built at Barrow and a long-term liability would be eliminated.

In summary, this new approach appears promising, but we believe more study is necessary on the MOD's part to determine feasibility and establish confidence in projected savings. Pending that, BAE Systems is moving forward with the proposal.

# CHAPTER EIGHT

# Conclusions and Recommendations

Our analysis indicates that fuelling all Astute-class submarines at DML (case 1) would be a highly undesirable option because of the high schedule risk it would incur, especially with respect to contractual hurdles, even though it should generate high savings by eliminating much of the effort related to anticipated site licensing costs at Barrow. Fuelling all boats beyond Astute 1 at DML (case 2) would realise modest savings, with a medium schedule risk. Waiting until Astute 4 to begin fuelling at DML (case 3) would not generate savings and would incur a comparable schedule risk as case 2. Compared with these cases, the original baseline has a lower schedule risk. It is important to note that our analyses of all DML fuelling cases assume that the Barrow site will be decommissioned, and they account for cleanup costs connected with past work related to low-level radioactive nuclear fuel at the site. After decommissioning, there would be no future regulatory costs related to handling nuclear fuel at Barrow.

BAE Systems has recently proposed to reduce nuclear fuel–handling hazards and resulting consequences at Barrow. These proposals show promise and should be investigated further for their risk-reduction measures and related savings. The proposals may provide the best balance of cost savings and schedule risk, but further analysis is required to confirm that.

The transportation challenges related to a fuelled as well as an unfuelled submarine exit out of Barrow need to be addressed immediately to avoid the risk of schedule delay to the Astute programme. If fuelling at DML is not considered for current and future Astute-class submarines, an unpowered exit would still be important if the option of power range testing at Faslane is to be given serious consideration. Such an option would require addressing the challenges related to transporting a fuelled submarine that has not gone critical and therefore cannot go under its own power out of Barrow to Faslane.

Based on these conclusions, we recommend that the MOD

- not consider fuelling the first Astute-class boat at DML
- take prompt action in analysing the latest proposal submitted by BAE Systems to reduce nuclear consequences of hazards at Barrow.

If upon further analysis the recent BAE Systems proposal is found unlikely to produce the savings and risk reduction anticipated, the MOD should engage with the regulators in assessing other options. It should look in detail at relevant aspects of build programme, support facilities, and options, and conduct a more detailed feasibility study for cases 2 and 3. The MOD should also consider the possibility that future nuclear regulatory requirements and restrictions could make both cases 2 and 3 seem advantageous, even with respect to the latest BAE Systems proposal. There is also the possibility, of course, that the current plan of fuelling all new boats at Barrow will emerge as preferable. Regardless, the MOD and BAE Systems need to

- review promptly the transportation challenges associated with moving Astute from Barrow to the open sea, regardless of whether the boat is fuelled or unfuelled, and produce a flexible exit plan that minimises potential schedule risk.

# TotalTide Measurement Methods

TotalTide is a specialist tide prediction programme published by the UK Hydrographic Office for Safety of Life at Sea (SOLAS). The UK Maritime and Coastguard Agency accepts TotalTide as meeting the requirement to carry tide tables under the Merchant Shipping (Safety of Navigation) Regulations 2002. It can be considered authoritative for predicting tides in the territorial waters of the United Kingdom.

The programme presents tidal information in a number of ways, and in this appendix we explain how we derived the data for our transportation analysis. We used the data for 2004; however, for navigational purposes, adjustments for the actual launch year will be required. TotalTide provides detailed tidal information at four points along Walney Channel: Barrow (Ramsden Dock), Haws Point, Roa Island, and Halfway Shoal.

For each tidal station, information can be displayed in a variety of ways, including tidal curves—we used this feature and the ability to measure directly from these curves extensively—and tide tables—we used these tables as a first filter to identify the curves we needed to investigate in greater detail. Details such as channel depth, vessel draught, and clearance were inputs to TotalTide.

Starting from the first tide in 2004, the full process for most of our tidal analysis followed these steps:

1. Input required safety parameters.

2. Determine safe heights of tide at Barrow (Ramsden Dock) and Halfway Shoal.
3. Inspect tidal tables to determine whether safe heights were achieved during that tidal cycle.
4. Measure rising tide for Barrow (Ramsden Dock).
5. Adjust safety parameters to meet lower channel specifications.
6. Measure falling tide for Halfway Shoal.
7. Adjust parameters for Haws Point and check curve.
8. Move to next tide.

For some of our work, we measured the time from high tide to the point of passing the safe height at Halfway Shoal. For these measurements, we took the tabulated time for high water at Barrow (Ramsden Dock) and measured to the falling point on the Halfway Shoal curve.

We took the vessel clearances described in the Vanguard Exit Plan[1] and input these in TotalTide as vessel characteristics. For channel depth, we used two sets of data: the actual depths reported by the harbourmaster,[2] and those from the Vanguard Exit Plan that are described as the channel design and are the minimum required by the maintenance dredging contract (see Table A.1).

## Detailed Analysis of Tides at Barrow-in-Furness

### Relationship Between Draught and Speed for Barrow Exit

We wanted to establish how the tidal system in Walney Channel would limit our options for transporting an unfuelled Astute from the dock system to safe water. Our initial research had given us an insight

---

[1] R. F. Hodge and M. Tansey, *Summary of Aspects of the Exit/Entry Project Relevant to the Transits of Walney Channel by VANGUARD Class Submarines,* Barrow-in-Furness, UK: Vickers Shipbuilding and Engineering Limited, June 1992.

[2] An Associated British Ports harbourmaster provided a locally produced chartlet with updated depth information.

**Table A.1**
**TotalTide Draught and Depth Parameters**

| Point | Parameter | Depth (metres) |
|---|---|---|
| Barrow (Ramsden Dock) | Exit plan minimum channel depth | 3.02 (rounded to 3.0) |
| | Vessel clearance | 1.5 |
| Halfway Shoal | Exit plan minimum channel depth | 5.08 (rounded to 5.1) |
| | Actual depth | 5.5 |
| | Vessel clearance | 3.5 |

to the challenges of fuelled departures from Barrow, specifically those of the Vanguard class of nuclear-powered submarines, and we wished to establish the boundary conditions. As we progressed with this work, we realised that we could extend our analysis of the relationship between draught and minimum speed to much shallower draughts.

We chose March 10, since visual inspection of the tide tables and curves showed that we would be able to get valid readings for vessels of deep draughts. Table A.2 shows our results using high water as the start time and the end point measured on the falling curve at Halfway Shoal. The channel depth used for Halfway Shoal was 5.5 metres, and the clearance required was 3.5 metres. Channel length was taken from the Vanguard Exit Plan. We did not account for the effect of a tidal stream in this analysis.

## Exit Opportunities

Our analysis of the relationship between draught and exit opportunities started with an assessment of the Vanguard Exit Plan. We used this as the starting point for our work because this plan is based on the dredging design for Walney Channel and because the Astute class has a similar draught to Vanguard's. We took a further step to adhere strictly to the exit conditions described in this plan, principally the safety margins and exit speeds, to better reflect the changed regulatory environment for the Astute.

**Table A.2**
**Draught and Speed Analyses Results**

March 10, 2004
Distance (nautical miles)    7.62
Barrow (Ramsden Dock)    HW 1335 at 9.6m

| Draught (metres) | Time (minutes) | Minimum Average Speed (knots) Assuming Departure at High Water |
|---|---|---|
| 5 | 237 | 1.9 |
| 5.5 | 220 | 2.1 |
| 6 | 204 | 2.2 |
| 6.5 | 189 | 2.4 |
| 7 | 173 | 2.6 |
| 7.5 | 157 | 2.9 |
| 8 | 140 | 3.3 |
| 8.5 | 125 | 3.7 |
| 9 | 114 | 4.0 |
| 9.5 | 93 | 4.9 |
| 10 | 75 | 6.1 |
| 10.5 | 53 | 8.6 |

NOTE: Speed (s) in knots has been calculated to one decimal place using distance (d) expressed in nautical miles and time (t) in minutes, according to the formula $s = (d/t)*60$.

Timing when the submarine starts the departure run is a key aspect for a fuelled departure. The submarine is aligned in the dock as the Ramsden Dock entrance is opened and with tug assistance uses its own power to pass through the entrance and turn to line up for the first leg. It needs to do this when the tidal stream across the entrance is minimal to nonexistent, and this time varies for each tide within the range of 45 to 30 minutes before high water.[3] We deducted 45 minutes from high water and used the later of either this time or the time measured from the rising tide curve as the start time for the departure. We then measured the time on the falling curve at Halfway Shoal and established the difference in hours and minutes. The Vanguard Exit Plan calls for a departure profile that takes 82 minutes, which is equivalent to 1 hour and 22 minutes. If the difference

---

[3] Vanguard Exit Plan and discussions with harbourmaster in March 2004.

we had measured was equal to or greater than this, we assessed the tide as feasible for a fuelled departure starting at 45 minutes before high water.

We reworked the start time for intervals of 40, 35, and 30 minutes to give an indication of feasible times across the comparatively slack tide departure window. The results are summarised in Table A.3.

To better inform our analysis of schedule risk, we investigated the relative impact of reducing the draught of the fuelled vessel and how it would affect exit opportunities. We applied the same procedures described above, using additional depths of 10.0, 9.5, and 9.0 metres, and undertook measurements for the month of February 2004 (see Table A.4). We chose February (which has 56 high tides) because it allowed the most exits for a draught of 10.3 metres, which is the maximum draught that can be accommodated within the Barrow dock system.

**Table A.3**
**Exit Opportunities, 2004**

| Start Minutes Before High Water | Number of Occasions per Year |
|:---:|:---:|
| 45 | 54 |
| 40 | 30 |
| 35 | 12 |
| 30 | 2 |

**Table A.4**
**Exit Opportunities, February 2004**

| Start Minutes Before High Water | Draught | | | |
|:---|:---:|:---:|:---:|:---:|
| | 10.3 m | 10.0 m | 9.5 m | 9.0 m |
| 45 | 4 | 13 | 29 | 38 |
| 40 | 2 | 8 | 28 | 38 |
| 35 | 2 | 7 | 25 | 37 |
| 30 | 2 | 4 | 24 | 36 |

# Bibliography

Buchanan, Henry, and Dan Wiper, *A2B Fuelling Study*, BAE Systems, ASTUTE Class Project, Reference 01/00/41000000/RP/8098026(5), ASTUTE Issue 3, 7 June 2002.

Cook, Cynthia R., John Schank, Robert Murphy, James Chiesa, John Birkler, and Hans Pung, *The United Kingdom's Nuclear Submarine Industrial Base, Volume 2: MOD Roles and Required Technical Resources*, Santa Monica, Calif., USA: RAND Corporation, MG-326/2-MOD, forthcoming.

Devonport Management Limited, *Options for the Initial Fuelling of ASTUTE Class Submarines at Devonport, Issue 1*, report provided to RAND on 28 April 2004.

Furness Enterprise, *UK Submarine Industrial Base: Employment Issues Associated with Nuclear Steam Raising Plant Installation in New Submarines*, presentation, 8 July 2004.

Hodge, R. F., and M. Tansey, *Summary of Aspects of the Exit/Entry Project Relevant to the Transits of Walney Channel by VANGUARD Class Submarines*, Barrow-in-Furness, UK: Vickers Shipbuilding and Engineering Limited, June 1992.

Nuclear Installations Inspectorate, and Chairman, Naval Nuclear Regulatory Panel, letter of understanding, 6 March 2003.

Schank, John F., Jessie Riposo, John Birkler, and James Chiesa, *The United Kingdom's Nuclear Submarine Industrial Base, Volume 1: Sustaining Design and Production Resources*, Santa Monica, Calif., USA: RAND Corporation, MG-326/1-MOD, 2005.

UK National Audit Office, *The Construction of Nuclear Submarine Facilities at Devonport*, Report by the Comptroller and Auditor General, HC 90 Session 2002–2003, 6 December 2002.